Human Being @ Risk

Philosophy of Engineering and Technology

VOLUME 12

For further volumes:
http://www.springer.com/series/8657

Mark Coeckelbergh

Human Being @ Risk

Enhancement, Technology, and the Evaluation of Vulnerability Transformations

Springer

Mark Coeckelbergh
University of Twente
Enschede, The Netherlands

ISSN 1879-7202 ISSN 1879-7210 (electronic)
ISBN 978-94-007-6024-0 ISBN 978-94-007-6025-7 (eBook)
DOI 10.1007/978-94-007-6025-7
Springer Dordrecht Heidelberg New York London

Library of Congress Control Number: 2012956296

Printed on acid-free paper

Springer is part of Springer Science+Business Media (www.springer.com)

Πάντα ῥεῖ

(aphorism attributed to
Heraclitus)

And what is purification but […] the release of the soul from the chains of the body?

(Plato, Phaedo)

Dying's a scandal. But we all do it.

(Don DeLillo, Cosmopolis)

In memory of Broos

Acknowledgments

I wish to thank everyone who contributed to the publication of this reflection on the intersection between ethics, philosophical anthropology, philosophy of technology, and philosophy of risk: the editors Pieter Vermaas and Evan Selinger, who encouraged me to continue this book project; Glen Miller, who helped me in preparing the final version of the manuscript by commenting on language and style issues; the anonymous reviewers of the previous versions of my manuscript; and my colleagues of the Department of Philosophy of the University of Twente, who provided me with an excellent research environment and gave me the opportunity to teach a course on the philosophy of human enhancement.

Contents

Chapter 1
Introduction

1.1 The Experience of Risk and Vulnerability

In his novel *Candide* (1759), Voltaire tells the story of a young man who hears from his teacher that this world is 'the best of all possible worlds' but then experiences a series of troubles and misfortunes that seems to challenge this optimism: Candide is flogged, witnesses a battle, learns that the girl he loves has been murdered, loses his protector in a storm, finds Lisbon destroyed by an earthquake, is beaten again, hears that his girl has been raped and used as a sex slave, must accept that she marries someone else for his money, has to flee from the inquisition, is almost eaten by cannibals, finds that the gold he just acquired is stolen from him, and when he finally settles into a more peaceful life, he grows bored. At the end, he finds happiness, but after having gone through so many troubles, that seems small consolation.

What happens to Candide is grotesque and is set in another age, for sure, but at the same time, it is very familiar to us. Although our lives might be less comic and less adventurous, they are not necessarily less tragic—or are at least *tragic enough*. Even if we are lucky and are now healthy; are surrounded by family, friends, and loved ones; are spared from financial problems; enjoy the work we do; and feel happy, we know that things can change for the worse at any time. We are not invulnerable gods. Vulnerability belongs to our existential condition.

As soon as we are born, and indeed even already *before* we are born, we are at risk. There is always the possibility that something bad will happen to us. Although the great majority of us do not experience the hardships of Voltaire's Candide or of other very unfortunate fictional characters, we know as a matter of *fact* that we are vulnerable to many bad things that can happen to a human being. We know that we are vulnerable beings. We know that we are highly dependent on the natural, social, economic, political, and technological environments we live in. We know that we can be hurt by other people or by a natural catastrophe, we know that we can have an accident, we know that there can be another financial and economic crisis, we know that our food can make us ill, we know that our personal computer and other electronic devices can be infected

M. Coeckelbergh, *Human Being @ Risk: Enhancement, Technology, and the Evaluation of Vulnerability Transformations*, Philosophy of Engineering and Technology 12,
DOI 10.1007/978-94-007-6025-7_1, © Springer Science+Business Media Dordrecht 2013

with a virus, we know that peace does not last forever, we know that there is such and such a probability that we get a life-threatening disease, and we know that we can die. And we know that eventually we will die. We know this. We see that bad things happen to people. We know that *one* does not live forever. We know the facts. Yet it is quite a different matter to become fully aware of these possibilities. What Heidegger says about death in *Being and Time* is applicable to risk and vulnerability in general:

> The publicness of everyday being-with-one-another "knows" death as a constantly occurring event, as a "case of death." Someone or another "dies," be it a neighbour or a stranger. People unknown to us "die" daily and hourly. "Death" is encountered as a familiar event occurring within the world. (...) One also dies at the end, but for now one is not involved. (Heidegger 1927, p. 234)

But this uninvolved 'objective' understanding of risk changes when something bad happens to someone we know or to ourselves. Then we might be liberated from what Heidegger calls the 'tranquillization about death' and experience 'Angst about death' (p. 235). When I experience *Angst*, I become aware of my '*ownmost* possibility' (p. 243): I realise that something bad can always happen to *me*, and death—*my* death—will happen *for sure*. In other words, I become aware of my existential vulnerability. I no longer think about risk in terms of probabilities; I am no longer lost 'in chance possibilities' (p. 243); I no longer think about risk and vulnerability as facts. I no longer say that 'one' is at risk or that 'human beings' are vulnerable. I acquire an *involved* understanding of risk and of the vulnerability of myself and of others. This is about risk that matters to me. It is about *my* vulnerability.

1.2 The Struggle Against Risk and Vulnerability

Even if we know the facts and even if we try to live with *Angst*, we have difficulties accepting our vulnerable existential condition. We revolt against it; we struggle against it. We struggle as individuals confronted with particular possibilities and particular sufferings, for example, when we struggle against a particular disease or when we refuse to accept particular bodily limitations, and we struggle against death.

We are like the young people in Kazuo Ishiguro's 'science-fiction' novel *Never Let Me Go* (see also the 2010 film), which tells a story about children 'bred' for organ donation; they will die young in the surgery room (the euphemism for this in the novel is 'to complete'). Towards the end of the story, the (by then) young people *know* that they have to die and usually accept this. Nevertheless, sometimes they struggle against their 'fate' and their increasing vulnerabilities, sometimes they rebel, and sometimes they try to extend their lifetime. Usually they protest silently, occasionally by means of argument and once by screaming. Their story seems to be a metaphor for our lives. Generally we accept death, especially death as a 'fact' of life, but when we grow older or when we get seriously ill, not all of us are happy to 'complete', and even if we accept that we will die one day, all of us struggle against human vulnerability; we try to reduce it as much as we can.

Partly, this unwillingness to accept the unavoidable is due to our human capacity of imagination: it is because we can *imagine* better worlds that we do not accept the view that we live in 'the best of all possible worlds'. It is also because of our imagination that we try to improve our condition and that we try to *create* a better world, for example, by means of technology (I shall return to this issue soon). Humans have always tried to reduce their vulnerability by developing and using technologies. In *this* sense, the rebellion against the human condition is neither an exclusive feature of the modern scientific project to make life artificial, as Hannah Arendt argued in *The Human Condition* (Arendt 1958, pp. 2–3), nor of the 'cybernetics' movement, as Dupuy argued (Dupuy 2008, 2009), but is part of what we are and what we do as humans.

However, the *modern* revolt may indeed be problematic in various ways. Consider for instance what Anders called 'promethean shame' (Anders 1956, pp. 23–25), a concept which we can interpret in this context as the shame that results from the realisation that the things we make are less vulnerable than us. Something like that might have happened to some fans of Steve Jobs. When the co-founder and former chairman and CEO of Apple died in 2011, it was experienced as a kind of 'scandal' by people who admired him: maybe because it made people aware of the sharp contrast between the shining, 'perfect', beautifully designed electronic products they used every day and the inescapable vulnerability and imperfection of their human, bodily existence. Looking into the mirror of our iPads and monitored by the calm electronic eye of our high-tech equipment, we might not feel 'zen' but instead ugly, imperfect, and vulnerable.

Promethean shame seems particularly relevant with regard to 'virtual' objects of 'cyberspace', which seem in principle to be immortal. Those who experience this kind of shame may wish—in Anders's words—to be 'made' rather than to be 'born'. Perhaps they want to be 'uploaded' to cyberspace in order to avoid the vulnerabilities related to bodily existence. Here Arendt's analysis of modern technology as an effort to leave the earth is relevant and compelling: modern technology seems indeed be motivated by a 'repudiation of an Earth who was the Mother of all living creatures under the sky' (Arendt 1958, p. 2) and perhaps by a desire to exchange the human 'for something he has made himself' (p. 3)—here we encounter Promethean shame again. (In Chap. 8, I will return to Arendt and to the point about leaving the earth.)

Arendt was mistaken, though, when she assumed that the earth provides human beings 'with a habitat in which they can move and breathe without effort and without artifice' and that there is no need for making life 'artificial' (Arendt 1958, p. 2). Instead, breathing, that is, our exchange with our environment, requires a far more active effort and involves artifice and technology so that human beings always make life 'artificial'—it is natural and artificial at the same time. This is so since we are artificial by nature, as Plessner argued (1928) and as contemporary philosophers of technology repeat (see e.g. De Mul and Verbeek): we need technology in order to live as humans. As the myth of Prometheus suggests, our natural instincts and our bodies are not enough; we are not complete, and we need technology (among other things). Thus, technology is part of our existential condition. In this sense, it is not external to the human. (I will return to this point in Chap. 2.)

Moreover, as I have said, we always struggle against our condition and try to improve it. Even if we do not experience Promethean shame, *we do not want to be vulnerable*, or at least not as vulnerable as we are (now). We resist. We rebel. We design and use technology to decrease our vulnerability. For example, we use medical knowledge and medical technology to prevent disease and postpone death. And we use all kinds of safety and security technologies to decrease our vulnerability. We clothe ourselves against the storms of misfortune. We protect and insure. We create safety architectures to hide from the Reaper, the Raper, and the Hacker. We live in institutional and technological bunkers that are supposed to shield us from misery and war. We are like Eric Packer in the film *Cosmopolis* (based on a novel by Don DeLillo): a young billionaire, who feels safe in his bulletproof stretch limo full of technology and wants to go to the hairdresser, is surrounded by death (a funeral), financial disaster, social protest, and threats to his life but nevertheless, continues to pursue his aim. This is an extreme case, but it is true that most of us go on with our lives and trust technologies *in spite of* risk. We too trust our safety cocoons and our panic rooms; we too try to hide in our own virtual world. We try to insure ourselves against the loss of our money, our autonomy, our control, and our health. We build material and electronic walls, fences, and dikes to keep out the viruses and dark waters of death. As technological beings, these are the sort of things we humans do. In fact, it is hard to imagine what our material culture would look like *without* the struggle against vulnerability: technology is our vulnerability guardian, and it is in the guardian's house that we live as technological, risk-phobic beings. We are vulnerable by nature, but we are also vulnerability-averse by nature. We are already rebels. We are the children of Prometheus.

Some people want to go one step further (or so it seems) and use technology not only to improve their individual condition or that of particular others but to *enhance* the *human*, to make us better-than-human. They consider the human malleable and improvable. They hope that human enhancement will make us less vulnerable. Some of them even desire to become completely invulnerable. Maybe they want the human to be *immortal*. But is this desirable? Is it *possible*?

In the 1950s, Arendt mentioned the possibility of making an artificial human; today, half a century later, new proposals for what is called 'human enhancement' have been made and have generated a lively and interesting discussion, partly prompted by new developments in biotechnology, information technology, and nanotechnology (sometimes called 'converging technologies'). Some say that 'playing God' is not a good idea or that human enhancement would amount to *hubris*: an insult to the gods that will be punished and thus end in disaster. Maybe human 'enhancement' creates more risks and vulnerabilities than it helps to mitigate. Others say that this is the way forward for humanity, that we should move on to a *posthuman* stage. And was not the desire for immortality always part of the religious aspirations of humankind, especially in the monotheistic religious traditions, which seem to have contributed to the birth of modernity and its Promethean project?

Whether or not we believe that 'human enhancement' is possible or desirable (I will discuss this question later), it is already worth noting that technology plays a double role here. Paradoxically, technology is at the same time the problem and the solution, the disease and the remedy: it helps us in our struggle against risk and vulnerability, but at the same time, technology also creates new risks. For example, in order to reduce our energy vulnerability, we have created technologies for oil production and for nuclear energy, but these technologies turned out to be sources of risk for humans as well as for the environment. We invented antibiotics and gave them to humans and to the animals we eat, but now we find out that we become vulnerable to antibiotic-resistant bacterial infections. This is paradoxical, and it will be a main purpose of this book to argue that new technologies—and, in general, new anti-risk measures and anti-vulnerability strategies—always create new risks and vulnerabilities, thus *transforming* human vulnerability rather than substantially reducing it.

Consider the historical context of technological development and technological risk. One hundred years ago, there was no risk of nuclear disaster; 50 years ago, there was no risk of computer viruses, let alone cybercrime; and 10 years ago, you could not be killed by a 'drone'. As our world changes as a result of new technologies, the risks we face also change. This also raises the question of whether human vulnerability, and indeed *the human*, remains the same, or whether it changes, depending on technology—among other things. I will argue in this book that the latter is the case and discuss its existential and ethical implications. But let me first begin to connect the previous discussion about human vulnerability to ongoing discussions about technology and enhancement in (applied) ethics and philosophy of risk. In particular, I will say more about the problem of technological risk as an ethical question and introduce the transhumanist argument for human enhancement.

1.3 Technological Risk and the Ethical Evaluation of New Technologies

It is by now nearly a platitude to say that new technologies create new opportunities but also involve new risks. However, usually we think we can *manage* these risks. But can we? Knowing, understanding, evaluating, and coping with these technological risks and understanding their ethical, social, and anthropological implications have turned out to be quite a challenge.

Consider, for instance, the risks associated with current nuclear technology, biotechnology, and information technology. Nuclear energy is a reliable way to produce energy for many users without using up carbon-based resources and without polluting the atmosphere, but it also creates radiation risks related to the disposal of nuclear waste and to the operation of nuclear power plants. They are never risk-free, as the recent nuclear disaster in Fukushima shows again. How can we deal with the ethical problems raised by nuclear waste disposal and radiation risk? Agricultural biotechnology promises to increase crop productivity and to minimise the use of herbicides and pesticides, but introducing genetically modified organisms may create risks for the environment. Are we prepared to pay that ethical price? The Internet gives us many benefits such as more information and worldwide communication at a relatively low cost, but it also makes possible new risks, such as the possibility of harm by computer viruses and cybercrime. How can we cope with these e-risks? And consider again the transhumanist project of 'human enhancement', which attempts to change human nature by means of information technologies and/or biotechnologies and which might therefore be considered as risky *par excellence*. What will happen if we 'play God'? Should we hinder or promote this project?

These questions are ethical but also political. New 'technological' risks and vulnerabilities may not only make us worry personally; given their ethical and social significance, they are also a matter of public and societal concern to such an extent that according to Beck, we live in a 'risk society' (Beck 1992): a society increasingly preoccupied with risk, in particular the risks produced by modernisation itself. As new technologies are invented and new artefacts are produced, new risks are invented, created, and produced. This has given rise to a kind of 'risk industry' that tries to study, assess, and control technological risk. In modernity this is our particular, *modern* way of coping with risk.

Note that this birth of new risks does not mean that so-called 'natural' risks have disappeared. 'Natural' disasters such as floods, earthquakes, and volcanic eruptions still happen. And, in spite of the transhumanist dream of invulnerability and immortality, we still become ill, we still age, and we still die. We remain vulnerable in this sense too, and not only individuals but also societies have to cope with that vulnerability.

However, even these so-called 'natural' risks depend on how we respond to them and are therefore not independent of new technologies. Rather, there is a complex relation between natural risk and technological risk. For example, as the climate change problem shows, we are highly dependent on our environment, but since we are not just passive recipients but also actively transform that environment, our 'environmental' vulnerability is as much 'natural' as it is 'technological'. Medicine and health care is another area in which we can see this entanglement of the natural and the technological. Medical technology and science reduce our vulnerability with regard to particular diseases by developing effective treatments. They even have significantly increased our life span. Yet these new technologies have also created new risks: consider, for example, the problem of antibiotic resistance or when new medical technologies become available and we have to renegotiate institutional arrangements in relation to medicine and health care and re-evaluate the distribution of (health) risks. For example, if it were to become possible for older women (e.g. age 70) to have children, should this be allowed and who should pay for the assisted conception? We also know more about individual risks because of new technology. For example, sequencing and analysing DNA of individuals seems to produce new information about diseases those individuals might (not) develop, although the status of that kind of 'knowledge' is still unclear. Again a 'risk industry' is created since society depends on experts who can assess those risks. Furthermore, even 'natural' risks can be reduced by means of technology, for example, by constructing dikes against flooding or by creating warning systems for tsunamis. If all this is true, then as individuals and as societies, we need to reflect not only on new technologies and the new risks and vulnerabilities they (may) create but also on their implications for 'old' risks and vulnerabilities, which are often reshaped by the new technologies and risks.

More generally, this preliminary discussion shows that it is unhelpful to separate discussions about 'natural' risk from discussions about 'technological' risk. Furthermore, based on the modern dualist assumption that 'technology' and 'the human' are two entirely different domains, philosophy of risk usually operates in isolation from philosophical reflection about the human. Much can be gained by bringing these two domains together.

Philosophy can contribute to how persons and societies cope with risk and vulnerability by reflecting on the meaning of risk and vulnerability, by reflecting on questions regarding the human (including questions concerning 'human enhancement'), and by contributing to the ongoing development of a framework for evaluating new technologies. This book attempts to do this by developing a normative anthropology of risk and vulnerability, which yields a framework that can simultaneously deal with the questions of what the human should be and which risks and vulnerabilities are acceptable. The conception of existential vulnerability that will be proposed will also be able to meaningfully integrate discussions about 'natural' and 'technological' risk; more generally, it will help to move beyond dualistic thinking about risk. Particular attention will be paid to ethical, political, and aesthetical issues raised by new information technologies and by proposals for human enhancement.

Note, however, that these arguments for a normative anthropology and for an ethics and politics of vulnerability are only relevant within the space for human agency that is available to us. It is a typical modern desire to want to bring everything under one's control, including the human and human vulnerability, but human being and human vulnerability are *not entirely in our hands*. Although we can shift the boundaries of control somewhat by means of technology, even that technological development is not entirely a matter of human agency: it is interwoven with what we may call techno-anthropological and techno-social *growth*. Therefore, in contrast to modern (scientific) belief, any project that seeks to alter human vulnerability can never *purely* be a matter of *design*. (I will return to this point in Chap. 2.)

1.4 Risk, Vulnerability, and Technology

What do we mean by 'risk'? There are many ways in which humans or non-humans are said to be at risk. For instance, we can say that we are at risk if we cross the street since we might get run over by a vehicle, that the ecosystem risks global warming, that we risk a nuclear disaster, etc. Risk concerns dangers (which can be great or small) and the possibility that bad things might happen. In our modern scientific culture, 'possibility' is interpreted in terms of 'probability'. Risk then refers to dangers and their probabilities (a high probability or a low probability). Thus, risk is about *what* might happen, not so much about *to* whom or what it might happen. A complementary concept, therefore, is vulnerability. This term emphasises the potential 'victim' of the risk, the entity that is at risk. A person or a thing is said to be vulnerable, that is, *at risk*. Since risk and vulnerability refer to different but related aspects of the same phenomenon—perhaps one could distinguish between its 'objective' aspect (the possibility of the dangerous, bad thing…) and its 'subjective' aspect (… happening to me)—I will use both concepts in my descriptive and normative discussion.

Yet this distinction may be confusing. As I already suggested in the beginning of this introduction, *both* concepts can be understood in an objective or subjective way. From a scientific point of view, risk is objective. We can try to measure it and adapt our designs and policies accordingly. It is seen as belonging to the realm of 'facts' and 'probabilities'. From an individual, human point of view, however, risk is always experienced risk—'perceived' risk, to put it in scientific, psychological terms. Experts doing risk assessment try to keep risk perception out of their assessment: it is subjective and therefore—in their eyes—less true or not true at all. Or risk perception itself becomes the object of scientific study: it becomes a 'bias' and a factor that is taken account in decision making and communication about risk. Similarly, vulnerability can be measured as an objective feature of things or bodies, but it can also refer to our subjective experience of being vulnerable. Which view is (more) true? Are the concepts of risk and vulnerability to be understood as objective states of affairs (facts) or subjective experiences or both? And how are both concepts related?

The approach offered in this book is that risk and vulnerability are neither subjective nor objective but are concepts that say something about the relation between subject and object. Let me pick up again my point about two aspects of the same phenomenon and summarise the existential-phenomenological turn I propose. First, what is missing from the objective conception of risk is *risk for whom or for what*. It is overly objective in the sense that it focuses too much on the external danger and leaves out the experience of risk by the subject. The concept of *being-at-risk* I will propose in this book is meant to close this gap by understanding risk and vulnerability in an existential-phenomenological way. This accounts for the 'subjective' aspect of risk and vulnerability mentioned before. Second, what is missing from the subjective conception of risk is the *object* that renders us vulnerable and *our relation to that object*. In so far as it remains dualist and non-relational, this conception renders risk disproportionately subjective in the sense that it suggests that risk and vulnerability are only 'in the head' or 'in the eye', a matter of 'mere perception'. But this denies the 'objective' aspect of risk and vulnerability. My interpretation of vulnerability in terms of intentionality, imagination, and transformation is meant to overcome these problems. By providing an existential (but not existentialist) analysis of risk and vulnerability, I argue that these concepts are best understood as being about the relation between subject and object, that is, as an existential condition and as an existential experience. Human being is always already at risk since we always already stand in relation and are engaged in the world. Both the 'objective' and the 'subjective' approaches to risk are particular ways of constituting the relation between subject and object, and both deny this existential condition and thereby deny the deep relationality of human being presupposed by the concepts of risk and vulnerability. Our relationality is a condition of possibility of risk experiences and practices. In other words, if we were as separate from the world as both views (objectivism and subjectivism) assume we are, then we would not be at risk and we would be invulnerable. We would be invulnerable since we would not be *engaged* in the world. Instead, we are always already engaged in the world, and this renders us vulnerable. I will argue that being-in-the-world is always a being-at-risk.

In modern philosophical, scientific, and technological praxis, however, it seems that invulnerability and disengagement is exactly what we aim for. Conceptually, we purify the entanglement of object and subject to such an extent that only a weak sense of relationality remains, if any. We sharply distinguish between objective and subjective, between perception and reality, and between me and the world. We acknowledge that there are relations, but these relations are merely external; we fail to recognise that subject and object mutually constitute one another and that there is an internal relation between them. Technologically, we now try to complete this modern philosophical project by material means: we attempt a Cartesian splitting of subject and object by using strategies of de-*world*ing and disembodiment. For example, some use information technology in order to try to escape to a virtual world. They try to separate mind from body. Others want to turn humans into invulnerable cyborgs with (partly) artificial bodies. Some hope that we can accomplish these projects in the future by using new, more advanced technologies. If such attempts were to be entirely successful, they would dehumanise us in the sense that they would destroy the specific human form of relationality and—as I will argue—the specific human form of vulnerability.

1.5 Transhumanism

Indeed, human being is also 'at risk' in the following way: some argue that in view of new technological developments, we face the emergence of the possibility of the disappearance of human being. With this phrase, I do not refer to natural or other disasters or catastrophes, global warming, wars, etc. that may destroy humanity but to the possibility of 'human enhancement' I already mentioned and in particular to *transhumanist* visions of human enhancement, which suggest the possibility of technological enhancements that move us beyond human being, make us *posthuman*, and which actively promote efforts that contribute to realising that possibility. Whether or not this is a real prospect, it is a powerful idea which deserves further philosophical attention.

In a FAQ document for the World Transhumanist Association (now Humanity+), Bostrom defines transhumanism as follows:

> Transhumanism is a way of thinking about the future that is based on the premise that the human species in its current form does not represent the end of our development but rather a comparatively early phase. We formally define it as follows: (1) The intellectual and cultural movement that affirms the possibility and desirability of fundamentally improving the human condition through applied reason, especially by developing and making widely available technologies to eliminate aging and to greatly enhance human intellectual, physical, and psychological capacities. (2) The study of the ramifications, promises, and potential dangers of technologies that will enable us to overcome fundamental human limitations, and the related study of the ethical matters involved in developing and using such technologies. (Bostrom 2003, p. 4)

Thus, an important transhumanist aim is to overcome 'fundamental human limitations'. The transhumanist project is partly motivated by the wish to eradicate, or at least radically diminish, human vulnerability. So-called posthumans are to be created which would be resistant to disease and ageing, have unlimited youth, and so on (Bostrom 2003, p. 5). Transhumanists hope (and sometimes predict) that by using technologies ranging from artificial intelligence to genetic engineering, we will reach this aim.

As I will explain in Chap. 2, this proposal has invited opposition from conservative thinkers who defend humanity as it is and argue that we are better served by accepting human vulnerability. Moreover, defenders and opponents of human enhancement differ on how to answer the question of what the human is. Essentialist conservatives tend to defend their view of human nature, whereas transhumanists argue that not only we have always changed (evolution) but also that we have always changed ourselves. Which of these views is right? What are we, and what should we (not) become? Should we try to overcome human being and human vulnerability? And *can* we become invulnerable? Are there limitations to 'making the human'?

Before I start my discussion of transhumanist human enhancement, let me summarise what I will do in this book.

1.6 Outline of the Book

In this essay on technology and vulnerability, I respond to discussions about risk and vulnerability and to discussions about radical human enhancement (as proposed by transhumanists) and new information technologies by sketching the contours of a normative anthropology of vulnerability, which asks what kind of vulnerability we want, given that our technologies already transform human vulnerability—and hence the human. I develop my argument in several steps. I start from the descriptive anthropological question and then move on to the normative question. Although the scope of my discussion is broad, my main concerns are ethical and anthropological, and my thematic focus is on the relation between vulnerability and technology. In the course of my argument, I critically respond to authors such as Bostrom and Habermas, but most of my effort goes into positive development: I outline a framework for a normative anthropology of vulnerability.

I offer a preliminary discussion only. This restriction is motivated by the following considerations: one is external and the other internal to the issue at hand. The first one is a matter of modesty. I regard the development of a full and comprehensive normative anthropology of vulnerability (and indeed a fully relational anthropology) as a collaborate task that goes beyond the limits of a monograph, beyond the capacities of an individual, and beyond the scope of a generation of scholars. Like other philosophical projects, normative anthropology does not happen at one place and has a past, a present, and a future. The second reason is epistemological and internal to the problem discussed. Since the issue of radical human enhancement concerns future technology and the future of human life, there are limitations about what we can say about it today. I will discuss this point in more detail as part of my argument in Part II.

Let me provide an overview of the content of this book. In the first chapter, I start with an overview of the discussion about human enhancement and formulate what I take to be the twofold transhumanist challenge in relation to the issue of vulnerability: the transhumanist vision questions traditional views about human being (an anthropological issue), and it questions traditional views about how to cope with human vulnerability (a vulnerability issue). I introduce the well-known controversy between transhumanists and conservatives (both bioconservatives and what I call 'infoconservatives') on the anthropological issue, but I also pay attention to a less discussed tension between those who accept vulnerability and those who want to overcome it.

Then I begin to respond to the anthropological issue. I take the edge off the novelty claim that clothes the discussion about human enhancement, argue for changing the object of the discussion from 'nature' to 'being', and clear the ground for a dynamic notion of human being that grants significant credit to theories that emphasise change, but without giving up the idea that there is something radically distinctive about what we are. First I argue that we must acknowledge that human nature has always evolved, that human being has always been changed by technology, that philosophical anthropology always had normative ambitions, and that the existential-phenomenological tradition offers a notion of human *being* that avoids the difficulties of both naturalist (biological) and essentialist (philosophical and religious) notions of human *nature*. I also stress, however, that accepting this dynamic notion of human being neither necessarily implies endorsing the transhumanist normative project nor accepting the assumption that the human is *entirely* open to design and that we can have *full* control over changes to the human. (I will accept that technology influences our vulnerability, but this is not the same as claiming that we can entirely *choose* what vulnerability we want.)

This preparatory work provides the basis for Chap. 3, in which I sketch some basic elements of an anthropology of vulnerability. First I outline what I take to be standard views of the relation between human being and risk: objectivist views, for example, in risk assessment, medicine, and the psychology of risk, as well as social-constructivist and cultural theories. Arguing against these dualist approaches to vulnerability, I propose to understand vulnerability as an existential condition. This approach allows me to say more about our specific way of being-at-risk and the role of imagination and emotion in being-at-risk. I discuss the relational-existential nature of vulnerability and point to some of the myriad ways in which vulnerability is created and sustained. Furthermore, I argue that a description of vulnerability is not exhausted by references to passive experiences of suffering but rather requires us to relate these experiences to what we *do* to render ourselves vulnerable. I also use Heidegger's vocabulary to develop my preliminary phenomenology of vulnerability. I end with a clarification of the relation between my existential-phenomenological approach and that of some others, which involves sympathising with the empirical anthropology of Jackson and distancing my approach from the *existentialist* dimension of Heidegger's (and Sartre's) view.

In Chap. 4, I first reflect on what I call 'cultures of vulnerability': I discuss the representations and imaginations of vulnerability that are constitutive of our vulnerability experiences, which in turn co-shape cultures of vulnerability. I provide the example of vulnerability related to illness and death. However, I also argue that culture is not just about representation and imagination but also about practices: we should also consider vulnerability *praxis* and vulnerability *habitus*.

In the next section, I further elaborate the relation between vulnerability and human action. I argue that our actions do not only render us vulnerable but also *transform* that vulnerability and therefore change what we are. I suggest a narrative about the development of vulnerability and culture, in which the struggle against vulnerability only tends to achieve the very opposite of what it intended: growing vulnerabilities and *new* vulnerabilities. For instance, we seek social security in communities and states, but this renders us vulnerable to violence of the group, the state, and the politicians who claim to represent us. And we seek material security in various kinds of anti-vulnerability architectures, but our technological, financial, and economic culture renders us vulnerable to new dangers and new forms of violence. We have tried to create more energy security but by doing that we have increased our environmental vulnerability. Our vulnerability has been *globalised*. I conclude that we have not diminished but transformed human vulnerability. Our societies, religions, and technologies have transformed our 'form of life' and hence have changed what we are. How can ethics respond to this challenge? This is my theme of the next three chapters.

In Chap. 5, I start the normative part of my inquiry: I explore what it could mean to engage in an 'ethics of vulnerability'. With regard to new technologies, I argue that, rather than starting from an essentialist notion of human nature or from a naturalist conception of human nature in order to approve or disapprove of particular technological developments, we first need to analyse the precise ways in which our vulnerability is transformed by technology, the ways in which we are at risk today. Then we can try to evaluate these transformations. However, I show how this approach differs from ethical theories that discuss, focus on, or even seem to *value* vulnerability, such as Christian ethics and ethics of care. Moreover, I suggest that ethics is not independent from technology: I question the means-ends dichotomy inherent in standard ethical theories (human values versus technological means) and propose an alternative non-dualist approach and noninstrumental view of technology, according to which both ethics and technology are part of how we try to understand and shape our vulnerability—the question is not *if* but *how* we want to be-at-risk. I also claim that in discussions about human enhancement, we should not only focus on potential transformations of vulnerability in the future ('designer babies', 'uploading', 'posthumans') but also consider how information and communication technology has *already* transformed what we are and *is* transforming what we are—a point I further develop in Chap. 7. Finally, I note again that the space for human agency with regard to vulnerability transformations is limited and that therefore an 'ethics of vulnerability' can guide our coping with technology, risk, and vulnerability only within these limits.

In Chap. 6, I first focus on the future of technology and further explore what the ethics of vulnerability can do as an 'ethics of human enhancement'. In the first section, I further reflect on the possibility and limits of such a project, given the possibility that our values may change as we change as a result of the new technological possibilities (see also the previous chapter). Can we predict radical vulnerability transformations and corresponding value transformations? Responding to insights from philosophers of technology (Jonas 1979; Anders 1956), I ask how we can imaginatively and emotionally cope with this challenge. How can we stretch our imaginative and emotional capacities for that purpose? In the next section, I illustrate my approach to this issue by briefly discussing ethics of information technology and ethics of genetic human enhancement in this light, and I explore how we can use literary fiction (e.g. Houellebecq's *La Possibilité d'une île*) to enhance our reflection. I end the chapter with an exploration of posthuman vulnerabilities. I argue that the transhumanist project must necessarily fail in so far as it aims to eradicate vulnerability and risk: if we were to become posthuman at all, we would still be vulnerable—albeit in different ways than now.

In Chap. 7, I return to the present and consider vulnerability transformations influenced by current information and communication technologies, which raise more urgent ethical questions. I explore the idea that contemporary information technology can be used as an anti-vulnerability tool. I acknowledge that it can be used to seek invulnerability and transcendence, but then I show that such a project necessarily fails. I argue that when we are using the Internet, we do not leave our bodies at home: although particular ways of 'online' experiencing and doing can give us the experience that we are removed from our bodies, from others, and from reality (e.g. I discuss the phenomenology of shooting), at a more fundamental level, our living and our experience remains embodied, social, and real, even if it is mediated by contemporary information technology. Therefore, when it comes to (evaluating) our present vulnerabilities, we should not make a sharp distinction between online and offline. Moreover, even if we could take distance from our biological bodies and from 'real' others, the 'virtual' world would not really make us invulnerable since it would create new vulnerabilities. I argue that to understand and deal with these new vulnerabilities, we need to go beyond dualist views of cyber security, which give us the false impression that we can retreat either into the autonomous castle of our hardware or into what we call 'reality'. As we have extended our vulnerability and become part of global ecologies of vulnerability, the inner/outer barrier vanishes, and full security is an illusion. Moreover, I show that the language of extension is misleading if it suggests that there was a human being separate from the environment in the first place; our vulnerability has always been ecological from the start. Finally, I discuss the relation between vulnerability and new technologies in medicine and health care. I argue that our present care-technological practices, though aimed at coping with vulnerability and at reducing vulnerability, create new dependencies and risks.

With regard to the discussion about human enhancement, these chapters on the ethics of vulnerability imply that, on the one hand, opponents of the transhumanist project do not sufficiently acknowledge and account for the variable nature of human being and its technological and social transformations. I propose to reframe the discussion: we should not ask whether or not human nature should be changed, but *which* changes we want and which changes we can imaginatively and emotionally cope with. On the other hand, these chapters suggest that transhumanists and other proponents of human enhancement should acknowledge that human enhancement technologies will not diminish vulnerability but create new vulnerabilities. Instead of dismissing opposition to their vision in terms of bias (Bostrom and Ord 2006), they should enter into a discussion about what kind of vulnerabilities are acceptable and what kind of vulnerable life we want to live.

However, this is not only, and perhaps not mainly, a question for individuals. How should we, as a society, respond to (the possibility of) new vulnerability transformations? In Chap. 8, I turn to the politics of vulnerability. First I distinguish between various meanings of this term. It can refer to political action that responds to violations of human rights, politics as an anti-vulnerability strategy, the way power is related to body and self, the threat of state violence, the distribution of vulnerability, (constraints on) anti-vulnerability struggles that harm others, the question what we as a society should do about vulnerability arising from human enhancement, the (evaluation of) political futures, and the redefinition—or perhaps reinvention—of politics, freedom, and democracy in the light of new technological possibilities.

Then I explore the relations between vulnerability and three well-known concepts in political philosophy: freedom, justice, and democracy. I start with freedom: I discuss the Hobbesian problem of social order in cyberspace and suggest a non-Hobbesian alternative approach, according to which cyberspace and its inhabitants are already social. Then I discuss the relation between justice and vulnerability. Taking up the distribution question introduced in the first section, I first explore a Rawlsian approach and ask the question concerning justice as fairness with regard to human enhancement: what could it mean to argue for a fair 'distribution of genes'? What does it mean to apply the difference principle to 'the distribution of genes'? For several reasons, I propose to discuss the issue as a problem of the fair distribution of *vulnerability* rather than genes. I also consider the possibility of using the concept of *capability* in relation to vulnerability and justice but argue that if we wish to think about capability transformations as transformations of vulnerability, we must reject the instrumentalist view of technology assumed by mainstream versions of the capability approach. I also argue that the question regarding a fair distribution of vulnerability cannot be restricted to the human world but must involve reflection on (our relations to) non-humans: we must ask the question regarding ecological justice for ecologies of vulnerability. Can we reimagine our world in a way that full acknowledges the connections between the vulnerability of humans and the vulnerability of non-humans? Finally, I explore the relations between democracy, freedom, privacy, and vulnerability. Responding to Arendt's view of political freedom, I ask if we should continue to conceptualise the public sphere as a sphere of invulnerability and reason. Are we prepared to accept that politics is always bio-politics and info-politics and that politics always involves vulnerability?

In Chap. 9, I take the last step in my normative inquiry by considering the aesthetic dimension of the normative-anthropological question. First, this is important with regard to the problem of Promethean shame: we want the human to be 'beautiful', to 'look good'. But what does this mean? How should (future) humans *look*? What *style* should we have? What is a *beautiful* design of the human? These are questions for a normative 'aesthetics of vulnerability'. I suggest and distinguish several interpretations and potential topics for discussion: art as a means to explore, give meaning, and reflect on existential vulnerability; how vulnerability looks, tastes, and smells, and how we *should* experience it. But the aesthetic dimension is also important if it is related to the ethical question in a more direct and fundamental way, at least if ethics itself is interpreted as a *technè* or craft: we can understand coping with human vulnerability as an art, which requires skill. Then there is an intrinsic relation between the ethics and aesthetics of vulnerability. I also ask the question whether humans should be the aesthetic 'measure of all things' and discuss the aesthetics of vulnerability with regard to the relation between humans and non-humans.

First, I show that we have always made representations of human vulnerability—vulnerability in the present but also vulnerability in the future. I argue that our present, modern stories are increasingly un-tragic and thus present a misleading picture of (in)vulnerability. I also briefly discuss the relation between spaces and vulnerability. I suggest that in late modernity, we try to create anti-vulnerability cocoons and anti-ugliness bubbles but that this does not diminish but merely transforms our vulnerability. I also argue that we always used animals, cyborgs, robots, and other (fictional or non-fictional) non-humans as 'mirrors' that helped us to define ourselves as humans, that helped us to delineate our human *form*. Furthermore, I develop the idea that coping with vulnerability is an art, a craft which requires skill. How can we become vulnerability artists? How can we cope with our vulnerability and that of others in such a way that beauty and good emerge? By asking these questions, we can weave together the ethical and the aesthetic threads of normative anthropology. Finally, I argue that existential vulnerability and human being are intrinsically ecological, linking us with non-humans, and that therefore their beauty (and their good) takes us beyond the 'merely' human. Emphasising again the limitations of what we can do to change human being and human vulnerability, I suggest that it is only partly up to us whether or not, and how, we continue the 'human song'.

References

Anders, Günter. 1956. *Die Antiquiertheit des Menschen* (*Volume I*): *Über die Seele im Zeitalter der zweiten industriellen Revolution*. München: Verlag C.H. Beck, 1987.

Arendt, Hannah. 1958. *The human condition*. Chicago: The University of Chicago Press.

Beck, Ulrich. 1992. *Risk society: Towards a new modernity.* Trans. M. Ritter. London: Sage Publications.

Bostrom, Nick. 2003. The transhumanist FAQ: A general introduction. World Transhumanist Association. Retrieved 6 Oct 2011 from http://www.transhumanism.org/resources/FAQv21.pdf.

Bostrom, Nick, and Toby Ord. 2006. The reversal test: Eliminating status quo bias in applied ethics. *Ethics* 116(July): 656–679.

Dupuy, Jean-Pierre. 2008. Cybernetics is an antihumanism: Advanced technologies and the rebellion against the human condition. *The Global Spiral* 9(3). Retrieved from http://anti-matters.org/articles/125/public/125-188-1-PB.pdf.

Dupuy, Jean-Piere. 2009. *On the origins of cognitive science: The mechanization of the mind*. Cambridge, MA/London: The MIT Press.

Heidegger, Martin. 1927. *Being and time: A translation of Sein und Zeit*. Trans. Joan Stambaugh. Albany: State University of New York Press, 1996.

Jonas, Hans. 1979. *Das Prinzip Verantwortung: Versuch einer Ethik für die technologische Zivilisation.* Frankfurt am Main: Insel Verlag. Trans. *The imperative of responsibility: In search of an ethics for the technological age.* Chicago: University of Chicago Press, 1984.

Plessner, Helmuth. 1928. *Die Stufen des Organischen und der Mensch: Einleitung in die philosophische Anthropologie* (Gesammelte Schriften IV). Frankfurt am Main: Suhrkamp, 1981.

Voltaire. 1759. *Candide, ou l'optimisme*. Trans. T. Cuffe. *Candide or optimism*. London: Penguin Classics, 2005.

Part I
Descriptive Anthropology of Vulnerability

Chapter 2
The Transhumanist Challenge

2.1 The Ethical Discussion About Human Enhancement and Its Assumptions About Human Being and Vulnerability

Philosophers have only interpreted human being, in various ways; the point, however, is to change it. This parody of Marx[1] could well be a slogan of the transhumanist movement,[2] which advocates radical ways of human enhancement. Human enhancement can be defined as the improvement of humans by technological means. There are many kinds of human enhancement, depending on the aim and the technology (means) proposed. For example, in a medical context, the term typically refers to improvements that go beyond mere therapy, such as improving the genetic make-up of an individual. But other technologies may be involved as well, and more likely a combination of technologies is used.[3]

In its radical version, the application of new technologies aims at moving us beyond human being towards new, posthuman modes of experience and existence. We are asked to welcome the possibility of enhanced memory, improved sensory capacities, and extended lifespans or even immortality. Once this will be realised in the future, so it is argued, why hold on to our old way of being? The time will soon come, it seems, when we can finally shed our clothes of human vulnerability and bravely step into the dawn of a radiant, posthuman future. Should we? Can we?

In the past decade, there has been much discussion about what may be called the normative ethics of human enhancement: is human enhancement morally acceptable, and if so, what kind of enhancement is morally acceptable? In this book, it is not my purpose to argue for a particular position in this debate. Rather, here and in Part II, I will engage with these issues in order to contribute to my main aim to elaborate what

[1] Marx famously asserted in his *Theses on Feuerbach* (Marx 1845): 'Philosophers have only interpreted the world, in various ways; the point is to change it' (thesis 11).

[2] In fact, Harris has referred to this phrase of Marx in his defence of human enhancement (Harris 2007, p. 3).

[3] Often the term 'converging technologies' is used (Nordmann 2004).

M. Coeckelbergh, *Human Being @ Risk: Enhancement, Technology, and the Evaluation of Vulnerability Transformations*, Philosophy of Engineering and Technology 12,
DOI 10.1007/978-94-007-6025-7_2, © Springer Science+Business Media Dordrecht 2013

I call *a normative anthropology of vulnerability*. In this chapter, I will respond to the controversy over human enhancement in order to introduce the issue of vulnerability and to start delineating my philosophical-anthropological position. In the next chapter, I will further develop my view by engaging with theory of (technological) risk.

2.1.1 Transhumanists Versus Bioconservatives and Infoconservatives: The Anthropological Issue

Before articulating my own approach, let me first examine the terms in which the current debate about human enhancement is framed. Transhumanist visions[4] have given rise to a highly polarised discussion between defenders of *human nature* and transhumanists, who want to change it. The latter accuse their opponents of irrational conservatism. For example, Bostrom and Ord claim that laypersons and others who oppose transhumanism suffer from *status quo bias* (Bostrom and Ord 2006). A further argument frequently used by proponents is that the development of technologies that will alter human being cannot be stopped anyway (Stock 2002). However, transhumanists are not always naïvely optimistic technology gurus such as Kurzweil (2005) or Naam (2005). Some recognise that there are potential dangers but argue that the proposed changes need not end up in a Brave New World. According to Agar (2004), a liberal eugenics is possible with respect for pluralism; Hughes (2004) defends a democratic transhumanism. Singer (2003) supports state intervention so that human enhancement is not left to the market, whereas Bailey (2005) thinks that we will learn by *trial and error*. Harris (2007) argues that even if we are bound to make mistakes, we still have a moral obligation to enhance and that choices of what to enhance should be left to the liberty of individuals. Opponents, by contrast, see great danger in human enhancement that should not just be regulated but forbidden. Consider the public outrage in Europe following Sloterdijk's proposal to make *rules for the human zoo* (Sloterdijk 1999): many people fear proposals for genetic enhancement, sometimes referred to as *eugenics*. Habermas (2001) sees a threat to human being and human dignity in the emergent biotechnological possibilities, as does Smith (2004); Fukuyama (2004) relies on the notion of human nature to argue in favour of therapy but against enhancement. McKibben (2003) sketches a horrific future and claims that our current technology is already *enough* that we should restrain ourselves and stop further developing the technologies. Elliott (2003) links human enhancement to a culture of obsessive happiness seeking by medical means. And Dupuy (2008) thinks we must 'defend humanism against the excesses of science and technology'.

The name *bioconservatives* does not cover all conservative responses to the issue of human enhancement, since other technologies, especially converging technologies,

[4] See, for example, 'The Transhumanist Declaration' from the World Transhumanist Association (2002), now Humanity+.

do not only belong to the *bio* domain. Therefore, let me add a new term to the vocabulary: next to bioconservatives, there are what I call *infoconservatives*, who do not want *information technologies* that change human being. Similarly, one could talk about *nanoconservatives*. (And of course, one could also avoid having to make these distinctions by talking about *the conservative response to human enhancement proposals*, regardless of which technologies they involve. However, I will continue to use the terms *bioconservatives* and *infoconservatives* in order to differentiate between the conservative responses and to indicate the significance of information technologies in relation to human enhancement.)

How can ethicists respond to the transhumanist project of radical human enhancement and to its critics, that is, to their *vision* (not necessarily the prospect) of the new human? And how can philosophy mitigate the polarisation in this discussion? First, we have to identify what is at stake. Terms such as *transhumanism*, *posthuman*, and *human nature* lead us to ask the question of philosophical anthropology: what is human being? What is this *human nature* opponents of transhumanism defend, and what is the *human* transhumanists wish to overcome? This link between ethics and anthropology is unsurprising, since conceptions of the good life are always related to conceptions of what we are. But it seems that now the question takes on a new dimension and a sense of urgency. The question is urgent, since if it is possible, at least to some extent, to change who we are by technological means, we better think about *what* we are changing. Moreover, it appears that the problem assumes a new dimension, a *normative* dimension: if and in so far we can change what we are, the anthropological question is not only what we are as humans (descriptive) but also what we should be (normative). It seems that for the first time in human history, philosophical anthropology becomes part of normative practical philosophy, since we now have the opportunity to change the human. Is this observation correct? Which method should a normative anthropology use? And how can this help us to defuse the tense controversy?

In the next sections, I will start to develop my contribution to answering these questions, but let me first say more about the relation between transhumanism and vulnerability.

2.1.2 *Fighting the Dragon or Accepting What Is Given by Nature or God? The Question Concerning Human Vulnerability and Technology*

The aims of transhumanism do not only challenge traditional views of human being; they also question our thinking about vulnerability. In order to sketch what is at stake in transhumanist visions of posthumanity, consider the following tale, told by one of the most prominent transhumanist thinkers:

> Once upon a time, the planet was tyrannized by a giant dragon. (…) It demanded from humankind a blood curdling tribute: to satisfy its enormous appetite, ten thousand men and women had to be delivered every evening at the onset of dark to the foot of the mountain where the dragon tyrant lived. (…) The misery inflicted by the dragon tyrant was incalculable.

> (...) Some people tried to fight the dragon, but whether they were brave or foolish was difficult to say. (...) Seeing that defeating the tyrant was impossible, humans had no choice but to obey its commands and pay the grisly tribute. (...) Spiritual men sought to comfort those who were afraid of being eaten by the dragon (which included almost everyone, although many denied it in public) by promising another life after death, a life that would be free from the dragon scourge. Other orators (...) said it was part of the very meaning of being human to end up in the dragon's stomach. (...) Most people tried to cope by not thinking about the grim end that awaited them. (Bostrom 2005, p. 273)

Clearly the issue addressed here is how we should cope with human suffering, ageing, and mortality—indeed with human vulnerability as an existential condition. For those who are brave enough to confront the problem, there are roughly two options. On the one hand, there are those who tell us we must accept all that is given to us by nature or God. Why not accept the outcomes of natural evolution or the will of God? Both *religious* and *nonreligious* humanists tend to take this position—the difference being that some religions promise an afterlife. Note that, in contrast to what Bostrom suggests in his tale, the traditional humanist position is not necessarily that we must accept *everything that happens to us* as a result of disease and ageing. Rather, traditional humanists urge us to reduce human suffering as much as we can, for example, by means of medicine. However, they accept *the rules of the game*, that is, they accept that being human involves disease, ageing, and morality. Transhumanists such as Bostrom, by contrast, are not content with this response: they want to change the rules of the game. They want to go beyond comforting and caring; they want to take away the root of the sufferings of humankind. This is the *moral* of Bostrom's tale: we should fight the dragon, that is, extend the (healthy) human lifespan, diminish suffering, and not accept ageing as a fact of life (Bostrom 2005, p. 277). In other words, when it comes to their final, long-term aim, transhumanists want to make us immortal, which means invulnerable. If this were possible, this would be the end of being human as we know it: we would become *post*human.

Transhumanists propose to use technology to reach this aim, through biotechnology (genetic enhancement) and information technology. For example, in *The Singularity is Near* (2005), Kurzweil suggests that we will become cyborgs, upload ourselves, have nanobots in our bloodstream, and enjoy nonbiological experience—whatever that means. Like many other transhumanist visions, these technological transformations seem to aim towards invulnerability and immortality: the idea is that we can transcend our present limited bodily existence by means of human enhancement technologies. Rather than being vulnerable mortals, we could become strong, invulnerable cyborgs or immortal minds living in an eternal, virtual world. *If* we had the means to bring this about, why would we refuse to kill the dragon of human vulnerability? Why would we refuse to become the gods we always feared and worshipped?

In the next chapters of this book, I argue that transhumanists can never reach their aim, at least if that aim is reaching absolute invulnerability. I argue that even if we had full control over the human and could fully *design* the human (an idea which is very problematic, see the end of this chapter) and enhance ourselves, become cyborgs, or upload our minds into virtual worlds or into machines, we would remain highly vulnerable entities. Transhumanist human enhancement would not erase our current vulnerabilities, but instead transform them. It would also create new vulnerabilities (see for instance Chap. 6). I conclude that posthumans also would have to cope with their own dragons. And that *today* we have to cope with the current vulnerability transformations, created by the growth of new technologies (e.g. information technologies) and with their human, social, economic, and political entanglements.

However, before I unfold and develop my argument about vulnerability in the next chapter, let me first respond to the philosophical-anthropological controversy that underlies the ethical discussion between conservatives and transhumanists. I will do this in a way that first sides with transhumanists but then begins to depart from *both* sides by articulating an existential-phenomenological approach to human being and human vulnerability.

2.2 First Response to the Anthropological Issue

Human enhancement technologies are new, emerging, or do not exist yet. In this sense, it is a *new* problem, raised by relatively recent technological developments and the visions and fantasies they kindle. But how novel are the stakes in the debate and its underlying anthropological dispute? Let me first diminish its aura of novelty by putting it in the context of three other earlier discussions: one about the variability of human nature, one about the relation between humans and technology, and one about what we are and should be. In addition, this contextualisation will help me to clarify my own approach to the anthropological problem. I will then prepare the ground for my own (positive) philosophical-anthropological inquiry in the next chapter by discussing what *kind* of philosophical-anthropological perspective could mitigate the polarisation in the discussion about ethics of human enhancement and help to open up new conceptual space for understanding and evaluating new technologies and the worlds and lives they shape.

2.2.1 Human Nature Has Always Changed

Until the nineteenth century, human nature was understood as invariable; it was viewed as a set of characteristics that defines the human as opposed to the non-human (animals, angels, demons, gods, etc.). From the mid-nineteenth century onwards, this conception of human nature was challenged by several theories that historicised human nature. But this had further consequences: the birth of the concept of history was also the death of the self-evidence of the very concept of human nature. First, biology[5] has put forward the idea of natural evolution: all organisms, including humans, change over time as a result of natural selection and (according to contemporary evolution theory) mutation and genetic drift. Second, the development

[5] Usually the name of Charles Darwin is associated with this view. Although Darwin is certainly not the only father of evolution theory, he is the most influential among them. Most contemporary evolutionary theories, therefore, are called neo-Darwinian. Today, the evolutionary approach has spread from biology to psychology and beyond.

of the social sciences[6] has produced a view of humans as heavily influenced—if not determined—by their social and cultural environment. Combined with the historical approach that emerged in the nineteenth century, these new insights suggested that today we are not the same *humans* than 10,000, 5,000, 2,000, or even 100 years ago. We have changed as biological and as social-cultural beings. In sum, during the last 150 years, the view has spread that there is no single and fixed human nature, a view that remains controversial today.[7]

Note that the implications of biological change for our view of human nature have received more attention than those of historical-social change. Note also that both kinds of theories can be interpreted in various ways, ranging from naturalism and determinism to interpretations that do not reduce what humans are and what they can do to their biological and/or cultural-historical aspects.

Referring to these developments in the history of ideas is helpful, since some of the contemporary resistance against human enhancement is still based on a pre-nineteenth-century view of who we are. It is important to understand what bioconservatives mean by *human nature*. Do they take into account relevant insights from evolutionary biology and the social sciences, and if so, how? A plausible view on the ethics of human enhancement should acknowledge biological and historical change. In particular, opponents to transhumanism should answer the question of why they think it is wrong to change human nature, given that there always has been evolution and given that our social and cultural environment has a significant impact on what we are. Consider, for example, how societal techniques to discipline people (e.g. Foucault 1977) change us. Consider how upbringing and education change us. They change what we are: not only who we are as persons in the course of our lifetime but also what we are as *humans* in the course of human history. Opponents should explain why they think using these techniques to enhance humans is right and human enhancement by (new) *technological* means wrong. Why, for instance, is genetic engineering wrong but (some forms of) social engineering right?

Let me further develop this point by discussing the arguments of an influential opponent of human enhancement. In *The Future of Human Nature* (2003), Habermas fears that the ethical self-understanding of the species will be changed in such a way 'that we may no longer see ourselves as ethically free and morally equal beings guided by norms and reasons' (Habermas 2003: 41). I share his fear for this development and sympathise with his moral ideal to some extent, but I think he is wrong if he suggests that human enhancement (his particular concern is genetic engineering) is the first blow to that self-image. The Kantian image Habermas relies on has been questioned by (among others) eighteenth-century *moral sentiment* thinkers such as Hume and Smith; by nineteenth-century giants such as Darwin, Marx, and Freud; and by twentieth-century positivist natural and social scientists and philosophers. Opponents to transhumanist human enhancement that rely on the Kantian ideal must explain why they think such enhancement would be a first or more fundamental

[6] Founding fathers of the social sciences include authors such as Auguste Comte, Émile Durkheim, and Karl Marx.

[7] Consider for instance opposition to (neo-)Darwinism by some religious groups or organisations.

blow to our self-image than ever before, why their moral ideal is more worthwhile to aim at than that of the transhumanists, why their own ideal does not *instrumentalise* persons (Habermas 2003, p. 44) given that in their view persons may appear as instruments in the hands of *pure reason*, and how much the transhumanist ideal differs from their own given that transhumanists may also appeal to *autonomy* and *reason* in order to defend its ideal: they may argue that human enhancement should make us more autonomous and more reasonable. Answering these questions would contribute to the *substantive* and *thick* discussion Habermas rightly seeks.

Let me respond to particular arguments made by Habermas to develop my point. First, his argument that *eugenetic interventions* bar the person from 'the spontaneous self-perception of being the undivided author of his own life' (Habermas 2003, p. 63) is problematic, since it assumes that without human enhancement we have such a high degree of autonomy, whereas this kind of autonomy—if it makes sense at all—is already challenged by social, cultural, and educational interventions, which shape us and make us at most a *co-author* of our own life. This does not render his argument pointless, but he and other opponents must (better) explain why the socialisation process is a smaller problem for our self-perception. I agree with the objection considered by Habermas in his postscript that 'the distinction between natural and social fate is less razor sharp than the way we usually understand it' (Habermas 2003, p. 83). Habermas's best response to the problem (earlier in the book) is that socialisation practice can be subjected to reappraisal by the person, whereas paternalistic genetic programming is irreversible (Habermas 2003, p. 64). But that argument does not acknowledge that our *natural* genetic make-up is also beyond our influence: it also has irreversible consequences and does not allow for reappraisal. As Habermas formulates the objection put forward by Nagel, McCarty, and others:

> Why should it make any difference for the moral person, within the network of her interpersonal relationships, whether her genetic inheritance depended on the vagaries of her parents' choice of partners and the world of nature, or from the decisions of a designer whose preferences are beyond her influence? (Habermas 2003, p. 81)

Consider again the young people in the novel *Never Let Me Go*: does it make a difference to the experience and meaning of their lives that they are clones and, in this sense, *designed*? The novel suggests that they experience similar interpersonal joys and difficulties and that (except for the issue that they are used for organs, which *does* make a significant *ethical* difference) the meaning they give to their lives is not radically different from the meaning we give to our lives, the beginning (and end) of which is also partly beyond our control. How relevant is the distinction between natural and artificial here? Habermas replies to this objection by saying that if we are unsure about the contingency of our natural roots, we may feel 'the lack of a mental precondition for coping with the moral expectation to take, even if only in retrospect, the *sole* responsibility for her own life' (Habermas 2003, p. 82). But surely we can never take the *sole* responsibility for our own life, given the many influences that shape it beyond our control. For example, our parents carry at least part of that responsibility, since they contributed to the *making* of our life by means of their education.

A better argument is to refer to the problem of equality (Habermas 2003, p. 115) or to say that the crucial difference lies in the possibility that in the former case we can hold someone accountable (our parents or other genetic *designers*), whereas in the latter, *natural* case, we cannot blame nature. In Habermas's words, 'The young person can call his designer to account, and demand a justification' (Habermas 2003, p. 82). This indeed alters our self-perception or identity (I see myself as someone who is designed by someone), but it does not necessarily change the perception of how *free* we are in relation to our genetic make-up. This aspect of our self-perception would only change if genetic engineering were available to us in the course of our lives (after birth), if we could redesign ourselves.[8] Moreover, technology can, could, and has been used to give us more freedom and autonomy. Habermas does not see this possibility. I agree with him that 'eugenic self-optimization of the species' is a 'horrifying prospect' if it is 'carried out via the aggregated preferences of consumers in the genetic supermarket' (Habermas 2003, pp. 92–93). We do not want *that* kind of autonomy. But technology need not be used in this way; there are other options. For example, biotechnology could be used to better enable persons to do their moral duty (to use a Kantian term) or to become better equipped to be a morally good person in other ways. I do not wish to argue for or against this option, but Habermas does not even consider alternative applications of biotechnology that may appear less *horrifying* to us—including those of us who sympathise with the Kantian view. Habermas's technological pessimism may be due to his well-known view (developed in his earlier work) that the *system* invades and colonises the *lifeworld* (Habermas 1981). If technology is understood as belonging to the *system* only and if it can only *colonise* our lifeworld, there is no room for seeing other relations between human being and technology.

In the discussion above, the question of how much control we do have and should have is a crucial one. Defenders of *human nature* do not only assume that human nature is fixed; they also assume a fixed border between culture and nature, between what is within our control and what escapes our control. But as Dworkin (1999) has argued, this distinction between *fate* and *choice*—part of the basis of our ethics of responsibility—is challenged by new biotechnological developments and is an important reason for our ethical worries. In other words, the border between culture and nature, between control and fate, is shifting or will likely shift in the future. This implies that in the discussion about human enhancement, we can no longer take this distinction as given and assume that there is consensus about it. The frontiers of control should be the object of explicit discussion. Moreover, if it is true that human nature has *never* been fixed and that there has always been an interactive relation between technology and human being, then the border between control and fate has never been fixed in the first place. We have always used technology to expand the domain of control, and, as I will argue below, we have always experienced limits to control. If this is true, human enhancement technologies do not by themselves pose

[8] Physicist Freeman Dyson suggests that in the future we will have available 'cheap and user-friendly tools and do-it-yourself kits for gardeners' and 'biotech games for children [...] played with real eggs and seeds' (Dyson 2006, p. 223). He says these tool kits should not be applied freely to humans. But if this vision became reality, who could stop parents design their babies? And who could stop adults redesigning themselves?

a categorically distinct ethical-anthropological problem as opposed to other technologies—although they make the problem more visible.

Let me now elaborate my point concerning the relation between technology and anthropology suggested in this section.

2.2.2 Technology Has Always Changed Who We Are

Historical change and the relationship between determinism and freedom mentioned above raise the question of how much influence we have on what we become because of social and cultural processes that operate outside of our control. Consider Marx's view. He is famous for what is taken to be his historical determinism but nevertheless left room for labour as conscious and imaginative design that exists outside of what is predetermined. In the first part of *Das Kapital* (1867), he suggests that what distinguishes us from animals such as bees and spiders, which are also capable of building, is our capacity of imagination (Marx 1867, Vol. I, Chap. 7). Moreover, in the same chapter, he points to the fundamental habit of humans to *transform* nature. This process of transformation he calls *labour*. We may also call it *technology*, the art of *technè*. Surely, Marx thought that the conditions under which people work are socially determined. But seeing the transformation of nature, that is, (the use of) technology, as fundamental for human nature, is an important step towards a richer conception of the human being. It implies that technological transformation is *natural*. This is also true for the transformation of human being. We changed ourselves because we are the kind of beings that change ourselves and the world by using technology; as I said in my introduction, this is part of our nature. It does not happen *in spite of* being natural. We are, as Plessner said, artificial by nature (Plessner 1928). Human self-transformations by technology can be observed in history, for example, in the history of medicine. Humans that live on average 80 years are no longer the same beings as those that lived on average 40 years. In this sense, our human nature has changed and is already changing. Even if one thinks that some (kind of) changes are bad (e.g. changes to lifespan), one has to accept the dynamic, natural-historical aspect of human nature. Furthermore, twentieth-century and contemporary philosophers of technology such as Ihde, Latour, Borgmann, and Verbeek (see e.g. Verbeek 2005) have studied the ways in which technology changed our existence by *mediating* the way we relate to the world and by shaping our relations with others. Whether or not it is a bridge too far to call us *cyborgs*, as Haraway (1991) did, in any case, the relation between humans and technology turns out to be more intimate than previously assumed.

For the discussion about human enhancement, this insight implies that opponents should not only acknowledge the influence of biological and historical change on who we are but also the influence of *technology* on our bodies, our relations, and our society—on who we are. This means that opponents of human enhancement are challenged to explain not only why they reject the purpose of human enhancement but also why this purpose and its means—the new technologies—are radically different from the purposes and technologies of the past, which influenced and shaped who we are. Is there a radical break with the past? For example, where is the line between *enhancement* and *therapy* in biomedicine, given that contemporary *therapies* have prolonged life in a significant way compared to the past? Habermas relies on a distinction between healing and enhancement (Habermas 2003, p. 44), but the *logic of healing* has also changed our lifespan—in this sense, the logic of healing is also the logic of human enhancement. Why is *that* prolonging of life good and the transhumanist goal of longevity and immortality bad? To put it in engineering terms, the difference between repairing and improving disappears (Cerqui 2002). Is this a problem, and why? Why is genetic engineering bad and previous biomedical technology good? Why would human enhancement technologies turn us into cyborgs, and previous technologies not? What exactly is wrong with seeing ourselves as cyborgs, if this means that we are shaped by technology?

To clarify my point, let me further comment on Habermas's arguments. He thinks that biotechnology creates *a new type of intervention*: what previously was given as nature 'now shifts to the realm of artefacts and their production' so that the boundary between what we are and what we make of ourselves disappears (Habermas 2003, p. 12). But if we have always been artificial to some extent, why is intervening in the human genome a *categorically new possibility*? And does it really raise moral questions 'of an *altogether different kind*' (Habermas 2003, p. 14)? Habermas is right if he says that the original philosophical question concerning the *good life* takes on new life, but he is mistaken in his view concerning the relation between what we are and what we make of ourselves. The border between the two has always been unclear. The *natural* human body has always been influenced by technology. Habermas rightly recommends us not to leave the discussion about the human to biologists and engineers, but by assuming a strict separation between human *nature* and *culture*—at least up to the point when we can intervene in the human genome—he subscribes to an impoverished and inadequate view of the human body, of technology, and of the relation between humans and technology. The 'categorical distinctions between the subjective and the objective and between the naturally grown and the made' (Habermas 2003, p. 42), between 'what is *manufactured* and what has *come to be by nature*' (Habermas 2003, p. 46), cannot be *uprooted* since they have never been as clearly differentiated as Habermas presumes. He may be right that the primary mode of experience is being a body rather than having a body, but this being a body, this experience of the lived body, is not isolated from technology and other material conditions. Rather, human being, including its bodily aspects, is shaped and transformed by technology. Inspired by existential-phenomenological strands in philosophy of technology, I will further develop this point below.

2.2.3 Philosophical Anthropology Has Always Been Normative

Philosophical thinking has always involved thinking about who we are, understood as *what the human is*. Today, fond as we moderns are of disciplines and subdisciplines, we call this 'philosophical anthropology' and differentiate it from *normative* subdisciplines such as ethics and political philosophy. But from Plato until today, these ideas have always had a significant *normative* side: ideas about who we are have always been coupled with ideas about who we *should* be. Long before transhumanists entered the stage, there was a strong current running through the history of philosophy—including philosophical anthropology—that was deeply dissatisfied with what we are and wanted to make us *better* than we are. In other words, it wants to change what we are.

Often the goal of these philosophers is moral improvement: we should lead better lives, become better people, and do the right things. We should become wise, reasonable, loving, self-controlled, etc. For this purpose, various techniques of moral education have been used. Although these techniques have always involved material aspects, we usually do not connect them to technological change. This should not surprise, since technology is often ignored as a theme in the history of moral philosophy and philosophical anthropology. For example, in Pojman's overview *Who Are We? Theories of Human Nature* (2006), the normative dimension of anthropology is present on virtually every page, but our relation to technology is simply not recognised as an issue. (We must turn to the still somewhat marginalised *philosophy of technology* to find such a discussion.) A more relevant work in this respect is Passmore's *The Perfectibility of Man* (1970), which traces the history of perfectibility from the Greeks to the twentieth century. In his first chapter, Passmore distinguishes between *technical* perfection (becoming good at performing a task) and teleological perfection (to reach an end, a *telos*—for Aristotle, the human *natural telos* is human flourishing, the good life, *eudaimonia*). And in Chapter10, he discusses *the perfecting of man by scientific progress*. However, his emphasis is on scientific and social change, not on technology as such. Science and technology are related, of course, but often (the material aspect of) technology and what it does to what the human is remain out of sight. Science is not only about *logic* but also about materiality. It is about labs but also about what humans do in their lives. It is *system*, but it is also *lifeworld*.

Thus, if *enhancement* broadly conceived is much more *mainstream* than many people suppose it is, defenders of human enhancement must explain how their project differs from other kinds of *enhancement* projects—historical or contemporary. For example, Bostrom has explained the difference between humanism and transhumanism: he understands transhumanism as a radicalisation of humanism[9]:

> Transhumanism has roots in secular humanist thinking, yet is more radical in that it promotes not only traditional means of improving human nature, such as education and cultural refinement, but also direct application of medicine and technology to overcome some of our basic biological limits. (Bostrom 2003a, p. 494)

---

[9] Note that there is also a deeper sense in which transhumanism can be understood as a kind of humanism. With Heidegger, we might trace the desire to change the human back to the modern humanist desire to gain control over everything that exists, to make everything a thing and a slave to our purposes. Dupuy, by contrast, understands 'cybernetics' as an antihumanism (Dupuy 2008). Although I agree with Dupuy that we should not reduce 'humankind to the status of an object that can be fashioned and shaped at will' or to a 'machine', I side with Heidegger in pointing out that this way of thinking is rooted in humanism, *or more precisely a particular kind of humanism*, which has been developing in modernity. I think this Heideggerian view provides us with a good argument against the 'transhumanist' human enhancement project. But criticising this particular way of doing and thinking neither prevents us from accepting some of the claims transhumanists make about the dynamic character of human nature and its relation to technology nor does it exclude the possibility

On the other hand, opponents of human enhancement should be able to explain why the inescapable *normative* aspect of the anthropological tradition they rely on (e.g. Platonism and Christian thought), in particular its aims and ideals, is different from the normative aims of transhumanism. If philosophical anthropology has always been *transhumanist* in the broader sense of wanting to improve the human, what is wrong with (contemporary) transhumanism? And even if we distinguish between the two (as I think we should[10]) and say that contemporary transhumanism is humanism by *technological* means, why is *that* wrong and humanism by educational means not? How different are there goals? And why are technological means wrong per se?

To put the challenge for the opponents of transhumanism in another way, if Kant would have had the *technological* means to overcome the biological *necessity* he struggled with in his philosophy and his life, would he have used them? *Should* he have used them?

The link between humanism and the transhumanist project of human enhancement has not always been recognised. For instance, in his controversial speech *Regeln für den Menschenpark* (1999), Sloterdijk has announced the end of humanism and has suggested a Platonic-Nietzschean project of selection by biotechnological means, which philosophy should reckon with. In making these claims and leaving aside whatever his own normative position is (which is not clear at all), he assumes a strong opposition between, on the one hand, a humanist culture which aims to *tame* people by education (reading literature, sitting, etc.) and, on the other hand, a biopolitical process of breeding and selection. Transhumanists, however, do not define their project as antihumanist but as humanist. They see it as humanism by other (technological) means (see again Bostrom 2003b). Thus, they do not see a sharp opposition between humanism and ethics, on the one hand, and a raw politics of power, on the other hand. Transhumanists can (and most of them do) consistently hold that ethics is still possible

---

of a different kind of humanism: a humanism that recognises and respects limits to human control and that develops a different kind of relation to the world, a less *modern* one. Given the 'technological nature' of humans, this does not necessarily mean that we must reject technology but rather that we need different kind of technology (and hence different kind of humans). And it seems to set up an ethical project rather than making such a project impossible, as Dupuy suggests (Dupuy 2009, pp. xviii–xix), although I fully agree with his Anders-inspired remark that currently we experience a gap between our capacity to act in the world and the ethical resources at our disposal. Moreover, we should not be blind to the *nonmodern* aspects of science and technology in practice. Perhaps things were always already 'out of control'; in contrast to what Dupuy, Arendt, and Anders seem to presuppose, we have never been all-powerful, pure masters. I suspect that even science and technology motivated by a 'transhumanist' human enhancement project will have this kind of ambiguity and paradoxicality. I will return to the issue of control at the end of this chapter.

[10] However, the distinction should not be made in terms of 'technological' versus 'non-technological' but in terms of the *kind* of technologies and objects used and—more importantly—in terms of the way they respond on the existential problem of vulnerability. Transhumanists try to reach immortality and invulnerability *on earth*, whereas Platonists and Christians tend to accept physical vulnerabilities and 'earthly' mortality. The soul may be immortal, but the body is not; death is the separation of soul and body. Even if the body is believed to be resurrected, this presupposes that the body first dies: one must fall to be raised but seek to transcend these vulnerabilities and to overcome death in a different sphere or in an afterlife (not on earth, not in this life).

and desirable and that we can and must make ethical choices concerning human enhancement that it is not—and should not be—merely a matter of power. For this reason, their view need not imply the brutal inequality usually connected to Platonic and Nietzschean elitist politics. Proponents of human enhancement differ on the political question: some want a liberal eugenics, and others argue for more state intervention. Some combine their view with a plea for inequality, and some defend equality. Thus, although not all humanists may defend (some forms of) human enhancement for other (ethical, political) reasons, there is nothing intrinsic about *humanism* broadly conceived that opposes it. Rather, many proposals for human enhancement are transhumanist without rejecting humanism, and it makes sense to define transhumanist human enhancement as a radicalisation of humanism, as Bostrom does.

Note also that, in contrast to what Sloterdijk thinks, new media need not be opposed to literature—the *old* technology and the humanist recipe for moral improvement—but can also be used as part of a humanist strategy. Computer games, for example, may assist Nussbaum's humanist mission if they meet certain ethical criteria (Coeckelbergh 2007). In general, technology and humanism need not be enemies. The only connection Sloterdijk sees between technology and humanism is the military one (a humanist discourse about universal rights that is employed together with a military intervention power), but fortunately this is not the only possibility for cooperation or symbiosis between the two.

2.2.4 From Human Nature to Human Being: From Essence to Existence

My purpose in the previous pages was not to side with either transhumanists or conservatives. Rather, my discussion so far suggests that both sides may be able to converge on a dynamic, variable notion of human nature while facing the task of defending their own view on whether or not human enhancement—the change of human nature by *technological* means—is allowed or should be encouraged.[11]

However, I am still dissatisfied with the notion of human *nature* that plays such an important role in the discussion as framed above. On the one hand, its use suggests too much the pre-nineteenth-century view I rejected as implausible and undesirable. On the other hand—and here I add further objections to transhumanism—if the notion of *nature* is understood in the terms of evolutionary biology only, there is the danger of naturalism, by which I mean the tendency to understand who we are *only* in the terms provided by natural science. (A similar danger lurks in deterministic social science.) In order to avoid these approaches, I would like to change the key term of the discussion from *human nature* to *human being*. Let me explain why; let me show what can be gained by this conceptual operation. This means that I will (have to) make explicit what kind of direction I believe we should take when thinking

[11] Note that later in this book I will further develop and modify my view (Chap. 6).

about what we are and which philosophical traditions can be helpful for this purpose. This will clear the ground for the anthropology I will develop in the next chapter.

How can we conceive of human nature in dynamic terms and still account for the intuition that human experience and human existence is unlike that of any other living being? What kind of approach to *the human* do we need?

First, we need a notion that allows for change but that also acknowledges stability. We are willing to accept evolutionary and historical change, but we still want to account for the intuition that there is *something* that is common to what it is to be human—something that remains stable over time. After the lessons of nineteenth- and twentieth-century natural and social sciences, however, we do not want to understand this *something* as a common *characteristic* or capacity such as reason, language, or intelligence; in short, we do not want to employ the notion of an *essence*.

Second, the social and psychological sciences have studied culture and (human) consciousness, but their (tendency towards) determinism fails to do justice to our intuition that we are, to a significant extent, *free* beings. (Note that in spite of the overt deterministic naturalism of some of its proponents, the transhumanist project assumes such a freedom, since it asks us to become aware of the possibility to shape our own nature.) Thus, we are willing to accept the influences of social and psychological processes on what we are, but we also want to leave room for human freedom—at least including the freedom to manipulate nature, to construct societies, and to shape ourselves (however problematic these conceptions of freedom and their consequences may be—see my discussion in the fourth chapter).

Third, we need a notion that acknowledges both the individual and social dimension of who and what we are. We experience the world from a first-person perspective, but this need not imply a view that disregards the importance of the social. We are *also* social beings; social relations and society co-shape who we are as individuals and what we are as humans.

Fourth, we need a notion that does not radically separate culture from biology, the human from technology, and culture from technology. The physical aspect of our being and the material aspect of the world we create(d) are both crucial for what we are as humans, and these aspects are deeply related to the social and cultural aspect of our world(s). As Latour has argued (Latour 1993), cultural anthropology, initially developed for studying non-Western cultures, provides us with a model of our own (Western world): we too live in a material culture. Latour has argued that there is a network of humans and non-humans; whereas modern thinking strictly separates humans and things, he argues that the social has always been a collective of humans and non-humans and that they have never been separated in the first place—and that in this sense we have never been modern. Whether or not we accept his claim that there is symmetry between humans and non-humans, this is a refreshing view of the social, which helps us to more fully appreciate the crucial, indispensable role of materiality and technology in human culture. And as I already mentioned, contemporary philosophers of technology also show how the human is very much intertwined with technology. For example, influenced by *postphenomenology*, Latour, and Borgmann, Verbeek has argued that material artefacts shape our existence and experiences: they mediate our experiences and our actions (Verbeek 2005).

I feel that the notion that responds best to these criteria is *human being* rather than *human nature* and the approach I endorse is inspired by existential phenomenology, pragmatism, existentialist philosophical anthropology, and contemporary philosophy of technology. Here I begin to offer my argument for this choice, but this argument will be continued in the next chapters: my analysis of vulnerability and its transformations will *show* the gains of this approach. Let me enumerate some methodological building blocks I borrow from the perspectives mentioned above.

First, an existential (but not necessarily existentialist) approach provides an attractive response to the problem of the variability of human *nature*. Instead of looking for the *essence* of human nature, it refocuses on the *existence* of human *beings*. Instead of asking what we are, it asks *how* we *exist*. With regard to human enhancement, this means that an ethics will have to rely on an anthropology that studies the potential influence of human enhancement technologies on human existence.

Second, phenomenology and pragmatism draw attention to the way we humans *experience* the world and the way the world *appears* to us. Whatever else it may be, what we call *freedom* is a particular experience we can have as humans: I experience a difference between me and the world, *and* it appears that what I do *matters* and changes the world (and vice versa). Freedom concerns my experience of the relation between me and the world. In that sense, freedom is unavoidable, as the existentialists (e.g. Sartre) knew. Thus, the claim about freedom is not understood as statement about the world but about human existential experience—how we relate to the world. We can acknowledge this in the discussion about human enhancement by not only discussing what the new technologies do to our genes, our body, etc., but also what they do to our existential experience. For instance, we can study whether or not, and in what ways, some (enhancement) technologies threaten or support our experience of freedom.

Note, however, that recognising this existential freedom and acknowledging that we have the possibility to change what we are do not necessarily imply that we must accept the (different) claim that the human is *entirely* a matter of technological *design*. First, existential freedom does not necessarily mean freedom of (technological or other) *action*. Rather, we are only committed to saying that humans have subjectivity and that we have *some* freedom of action. Let me start with the former. Sartre writes:

> For we mean to say that man primarily exists – that man is, before all else, something which propels itself towards a future and is aware that it is doing so. Man is, indeed, a project which possesses a subjective life, instead of being a kind of moss, or a fungus or a cauliflower. (Sartre 1948, p. 28)

This is important for vulnerability: as humans, we are *aware* of our vulnerability, and as I will argue, part of human vulnerability is created by our particular mode of experiencing vulnerability. However, recognising this existential freedom does not necessarily imply that we have full freedom of action, let alone technological action. The space within which we can *make* or *design* the human is subject to various constraints, and once we create or design something within that space, its ethical and social consequences and meaning are not restricted to our concept and design. We can think about what we want when it comes to vulnerabilities, but we cannot necessarily *have* what we want. This is especially true for specific vulnerabilities, which are not just up to design. For example, if we want to engage in *genetic engineering*, we do not have full control over the human genome since we do not fully know what all the genes do, and once we change something, we do not fully control the biological, ethical, and social consequences of this *genetic* change. But this is not only so in the case of *human enhancement* technologies; all technological change is only partly a matter of design. To think that we can—in principle—fully control and design the world and that we can become masters of nature and masters of society is a typical modern error. To think that we can fully master technology is a similar error. Heidegger's term *fate* is maybe too strong a term, but his reflections on technology (e.g. Heidegger 1977) suggest that the emphasis on control and the means-end thinking that goes with it is part of the problem of modern technology. Technology and vulnerability are only partly the result of planning and control; like the human, they come into being and grow within a particular environment, which shapes them and which is shaped by them. Technology shapes our ways of thinking and doing as much as we shape technology. And what vulnerabilities we will have depends only partly on us. Hence if I talk about designing the human, about a *normative anthropology*, and about an ethics of human vulnerability, this should always be read with the qualification *in so far as* or *to the extent that* we can design and *in so far as* or *to the extent that* we can control the form of our vulnerability and what the human is. Those who employ the sorcerer's apprentice metaphor and suggest that first we had mastery over technology and then lost control (e.g. Dupuy 2009, p. xiii) fail to take into account that both technological action and the human have always been somewhat *out of control* and that both have always been partly designed and partly *not* designed—that is, the human is always also a *given*.

Third, the human existential condition is deeply social and cultural. All serious reflections in philosophical anthropology have acknowledged this, but it remains difficult to understand the precise relation between the individual and the social-cultural. Cultural anthropology tends to put too much emphasis on the social-cultural side, whereas most existential accounts tend to over-emphasise individual experience. In any case, the notion of human being leaves more conceptual space for the social aspect of human existence than the concept of *human nature* did, at least if the latter is understood as referring to features of humans as individual organisms. With regard to human enhancement, it is important to keep the social-cultural aspect in mind. The ethical-anthropological question concerning human enhancement is not (only) about what technology may do to individual existence but also about what it may do to society and to communal existence. Philosophical anthropology can be aided by cultural anthropology for this purpose. For example, as I will show in the next chapter, we might take inspiration from empirical, narrative *existential anthropology* (Jackson 2005), which connects existential phenomenology with cultural anthropology, pragmatism, and critical theory.

Fourth, as I already said several times, human being should not be understood as radically separate from technology. Modern technology may be problematic. But changing and shaping matter and bodies—even if that always takes place within particular constraints and can never be fully controlled—is part of what human existence is about. The notion of *human being* is broad enough to include this technological aspect of human existence; whereas the notion of *human nature* suggests too much separation between *nature* and technology, between *nature* and culture, and between *natural* humans and the technological world. An anthropological and ethical analysis of human enhancement technologies should employ a notion of human being that connects human being with technology and clarify this connection. Contemporary philosophy of technology can help to achieve this aim.

Having explained my choice of the term *human being* and my approach, let me introduce some advantages of my focus on vulnerability as the main theme of this book. These advantages both strengthen and complement those I associated with the term *human being*. The concept of vulnerability I will employ in the next chapters has the advantages that it connects facts and feeling, the objective and the subjective, body and mind, sense and perception, and world and experience (see especially my arguments in the next chapter). Vulnerability can be framed in the language of natural science (in probabilistic rather than deterministic terms, given the nature of risk), but it can also refer to our existential experience of vulnerable being. Used in the latter way and used in conjunction with the term *human being* and the existential approach, the term vulnerability allows us not to reject science and its naturalistic approach to human being but to argue for an approach that goes beyond a mere scientific analysis of what it is to be human. It is a broader approach that reveals the scientific view as only one way of framing human being and human vulnerability. Furthermore, I will understand human vulnerability as variable, as changing, which contributes to understanding human being as variable. How it may change is partly in our hands. In order to exercise our responsibility with regard to that handling, we need a descriptive and normative inquiry into vulnerability changes—and therefore changes to the human.

In the next chapters, I will sketch an anthropology of vulnerability, which is to inform an ethics and politics of technology (present technology and future technology), including an ethics and politics of human enhancement and of new information technologies. I will start with a descriptive anthropology, and then I will turn to an outline of elements for a normative anthropology of vulnerability.

References

Agar, Nicholas. 2004. *Liberal eugenics: In defence of human enhancement*. Malden: Blackwell.

Bailey, Ronald. 2005. *Liberation biology*. Amherst: Prometheus.

Bostrom, Nick. 2003a. The transhumanist FAQ: A general introduction. World Transhumanist Association. Retrieved 6 Oct 2011 from http://www.transhumanism.org/resources/FAQv21.pdf.

Bostrom, Nick. 2003b. Human genetic enhancements: A transhumanist perspective. *The Journal of Value Inquiry* 37: 493–506.

Bostrom, Nick. 2005. The fable of the dragon tyrant. *Journal of Medical Ethics* 31(5): 273–277.

Bostrom, Nick, and Toby Ord. 2006. The reversal test: Eliminating status quo bias in applied ethics. *Ethics* 116(July): 656–679.

Cerqui, Daniela. 2002. The future of humankind in the era of human and computer hybridization: An anthropological analysis. *Ethics and Information Technology* 4: 101–108.

Coeckelbergh, Mark. 2007. Violent E-games, empathy, and cosmopolitanism. *Ethics and Information Technology* 9(3): 219–231.

Dupuy, Jean-Pierre. 2008. Cybernetics is an antihumanism: Advanced technologies and the rebellion against the human condition. *The Global Spiral* 9(3). Retrieved from http://anti-matters.org/articles/125/public/125-188-1-PB.pdf.

Dupuy, Jean-Piere. 2009. *On the origins of cognitive science: The mechanization of the mind*. Cambridge, MA/London: The MIT Press.
Dworkin, Ronald. 1999. Playing god. *Prospect Magazine*, May 20, 1999.
Dyson, Freeman J. 2006. The domestication of biotechnology. In *What is your dangerous idea?* ed. John Brockman. London: Simon & Schuster UK/Pocket Books, 2007.
Elliott, Carl. 2003. *Better than well: American medicine meets the American dream*. New York/London: Norton & Company.
Foucault, Michel. 1977. *Discipline and punish: The birth of the prison*. Harmondsworth: Penguin.
Fukuyama, Francis. 2004. Transhumanism. *Foreign Policy*. Retreived from: www.foreignpolicy.com/story/files/story2667.
Habermas, Jürgen. 1981. *Theorie des kommunikativen Handelns*. Frankfurt am Main: Suhrkamp.
Habermas, Jürgen. 2001. *Die Zukunft der menschlichen Natur: Auf dem Weg zu einer Liberalen Eugenik*. Trans. H. Beister and W. Rehg. *The future of human nature*. London: Polity Press, 2003.
Habermas, Jürgen. 2003. *The future of human nature*. Cambridge: Polity Press.
Haraway, Donna. 1991. A cyborg manifesto: Science, technology, and socialist- feminism in the late twentieth century. In *Simians, cyborgs and women: The reinvention of nature*, 149–181. New York: Routledge.
Harris, John. 2007. *Enhancing evolution: The ethical case for making better people*. Princeton: Princeton University Press.
Heidegger, Martin. 1977. The question concerning technology. In *The question concerning technology and other essays*, Heidegger, Martin, ed., trans. W. Lovitt. New York: Harper & Row.
Hughes, James. 2004. *Citizen cyborg: Why democratic societies must respond to the redesigned human of the future*. Boulder: Westview Press.
Jackson, Michael. 2005. *Existential anthropology: Events, exigencies, and events*. New York/Oxford: Berghahn.
Kurzweil, Ray. 2005. *The singularity is near: When humans transcend biology*. New York: Viking/Penguin.
Latour, Bruno. 1993. *We have never been modern*. Trans. C. Porter. Cambridge, MA: Harvard University Press.
Marx, Karl. 1845. Theses on Feuerbach. In *The German ideology*. New York: Prometheus Books, 1998.
Marx, K. 1867. *Das Kapital: Kritik der politischen Ökonomie* (Part I), In *Werke*, ed. K. Marx and F. Engels. Berlin: Dietz Verlag, 1962.
McKibben, Bill. 2003. *Enough: Staying human in an engineered age*. New York: Times Books.
Naam, Ramez. 2005. *More than human*. New York: Broadway Books (Random House).
Nordmann, Alfred (ed.). 2004. *Converging technologies: Shaping the future of European societies*. Brussels: European Commission.
Passmore, John. 1970. *The perfectibility of man*. Indianapolis: Liberty Fund, 2000.
Plessner, Helmuth. 1928. *Die Stufen des Organischen und der Mensch: Einleitung in die philosophische Anthropologie* (Gesammelte Schriften IV). Frankfurt am Main: Suhrkamp, 1981.
Pojman, Louis P. 2006. *Who are we? Theories of human nature*. New York/Oxford: Oxford University Press.
Sartre, Jean-Paul. 1948. *Existentialism and humanism*. Trans. P. Mairet. London: Methuen & Co.
Singer, Peter. 2003. Shopping at the genetic supermarket. Retrieved from: http://www.petersingerlinks.com/supermarket.htm.
Sloterdijk, Peter. 1999. *Regeln für den Menschenpark*. Frankfurt am Main: Suhrkamp Verlag. Trans. P. Beers. *Regels voor het mensenpark: Kroniek van een debat*. Boom: Amsterdam, 2000.
Smith, Wesley J. 2004. *Consumer's guide to a brave new world*. San Francisco: Encounter Books.
Stock, Gregory. 2002. *Redesigning humans: Our inevitable genetic future*. Boston: Houghton Mifflin Company.
Verbeek, Peter-Paul. 2005. *What things do: Philosophical reflections on technology, agency, and design*. University Park: Pennsylvania State University Press.
World Transhumanist Association. 2002. The transhumanist declaration. Retrieved from http://transhumanism.org/index.php/WTA/declaration/.

Chapter 3
Anthropology of Vulnerability

3.1 Standard Dualist Views of Risk and Vulnerability

In order to outline an anthropology of existential vulnerability, I first need to distinguish my approach from existing approaches to risk and vulnerability. In this chapter, I start with what I consider *objectivist* views of risk, which presuppose modern dualist thinking. This will help me to define an existential-phenomenological view of risk and vulnerability, which will be further refined by engaging with the work of Heidegger and Plessner.

3.1.1 Objectivist Views: Risk Science, Medicine, and the Psychology of Risk

The standard approach to risk and vulnerability is an objectivist one. Risk and vulnerability are facts about the world that can be calculated, assessed, and managed. Experts try to calculate probabilities and assess consequences. They try to calculate financial loss and public health risks (e.g. cancer risk as a result of specific materials). They try to assess the risk of tsunamis, and they try to calculate the vulnerability of a technological system such as an oil production facility, a nuclear power plant, or an information and communication system. So-called *vulnerability assessments* are made in order quantify the vulnerabilities of information technology systems, energy supply systems, water systems, and so on. The experts monitor, manage, and reduce risk. On this view, it does not matter how humans feel about the risk. Humans and their communities are taken into account, of course (in particular, in *social vulnerability* studies), but usually they are seen as study objects that can be impacted by a hazard. They are not regarded as subjects. And *if* their experience is considered at all, it is often assumed that experts know the real risks, whereas the public responds *emotionally* out of ignorance. What we need, on this view, is proper

M. Coeckelbergh, *Human Being @ Risk: Enhancement, Technology, and the Evaluation of Vulnerability Transformations*, Philosophy of Engineering and Technology 12,
DOI 10.1007/978-94-007-6025-7_3, © Springer Science+Business Media Dordrecht 2013

risk communication: educate the public about the objective risk, and in this way avoid opposition to the particular technology. Moreover, if a disaster happens, it is seen as the result of *bad management*. There is little or no sense of what we may call *fate* or *tragedy* (see Chap. 9). Even the earth is seen as something we can and must manage and control: climate change becomes a management problem, and the earth becomes real estate; it is our property and we can buy, sell, use, manage, and rebuild it. Within modern thinking, there is only space for control, mastery, and management. It is assumed that the world is in our hands. The world is a *standing reserve* (Heidegger 1977) of resources which we use for our purposes, and the management of risk and vulnerability is part of that game.

Similarly, the medical sciences understand human vulnerability as the (sum of the) objective vulnerabilities of the human body. Their purpose is not only to intervene when bad things happen but also to study the vulnerabilities of the human body and, if necessary, intervene in order to control and reduce health risks related to the sick body (prevent worse things from happening) or the healthy body (prevention). What matters to medicine (understood as an objective science) is not how humans experience these vulnerabilities and risks but what the *real risks* are. According to this view, patients should be educated about the risks related to their specific bodily vulnerabilities and their lifestyles. For example, smokers are told that they risk developing lung cancer if they continue smoking. People are told to wash their hands regularly in order to reduce the risk of bacterial infections. Measures can then be taken to reduce specific vulnerabilities and risks. Furthermore, health is not seen as something that depends on chance, luck, or gift but something we can manage. Health is supposed to be managed by the state, by insurances and other health organisations, and (increasingly) by people themselves. We are supposed to act as the managers and stewards of our health and our bodies.

Not all scientists neglect how people experience risk. Another approach to risk is to psychologise risk and study individual risk perceptions. People are asked what they feel about a risk. However, it is questionable if this really is a different approach at all. Even if psychologists of risk do not discount subjective experience (e.g. Slovic 2000), *objective* risk assessment is contrasted with subjective risk *perception*. This terminological scheme suggests that there is an objective, real risk by which subjective experience can be measured and judged (Coeckelbergh 2008). Thus, psychologising risk is not a real alternative to the objectification of risk but presupposes it. In fact, this approach increases the scope of objectification: the subjective perception is also objectified. Experiences of risk and vulnerability are turned into statistics and data. It is depersonalised and detached from the context in order to manage risk. Are there real alternatives to the objectivist position? Consider the following options.

First, we could say that risk and vulnerability belong to human nature. We are, as MacIntyre argues (1999), dependent animals, and this dependence makes us vulnerable. Similar arguments can be found in the so-called ethics of care literature. But in so far as this literature neglects how people experience risk and remains blind to the phenomenology of vulnerability, it seems that this view is only objectivism in disguise: vulnerability appears as an objective, essential feature of human nature, and the vulnerability of people is studied in an objectivist way. To the extent that neo-Aristotelian and ethics of care perspectives on human vulnerability are essentialist and modern, they tend to view humans as objects with certain essential characteristics. We will need to take an existential turn in order to move away from that view. (More discussion of MacIntyre follows in Chap. 5).

Second, one could claim that risk and vulnerability are entirely *subjective*: it is about how *I* experience risk and about how *I* experience *my* vulnerability. However, such a view neglects that my experiences of risk and vulnerability are always directed towards an object that co-constitutes the risk and that renders me vulnerable—I call this the *intentionality* of risk and vulnerability as existential conditions—and therefore this view remains stuck in the dualist objective-subjective conceptual scheme. I will develop a view that goes beyond subjective/objective dualism.

Third, one could understand vulnerability in social-contextual or environmental terms. This makes sense, since vulnerability seems to depend on social and cultural arrangements. For example, a disability in one culture or situation may be an advantage in another culture or situation. As the saying goes, in the land of the blind, the one-eyed man is king. What is considered to be risky and what kinds of risks are produced depend on one's society and one's culture. Risk and vulnerability are relative to the social arrangements we put in place. Moreover, from a naturalist point of view, the vulnerability of an organism depends on the (changing) environment of that organism. A fish is very vulnerable on the land, but not in the water (and for us the opposite is true). Risk, on this view, is a feature of the natural environment, not of the organism. But whether we take a social-contextual or natural-environmental perspective, it turns out that in these approaches risk and vulnerability appear as objective once again: they are viewed as features of the social-cultural situation or of the natural environment. From a sociological or biological point of view, the *experience* of risk and vulnerability is not relevant.

3.1.2 The Social Construction of Risk and Cultural Theory of Risk

In order to further show why sociological and cultural views are (still) objectivist, let me briefly discuss the concept *the social construction of risk* and the so-called cultural theory of risk.

The social construction of risk concept can be understood as a response to the methodological individualism of risk psychology, which treats individuals as if they are separate from their social and cultural context. Moreover, it is also a response to the objectification of risk by the natural sciences, which obscures the fact that risk also depends on social agreements and social structures. In this sense, risk is socially produced, that is, produced by social interactions. This means that risk is not only *sociologised* but also historicised. For example, Beck has argued that the production, identification, and assessment of risk depend on the kind of society in which we live, which now happens to be modern *risk society* (Beck 1992). He has claimed that late modernity is characterised by *simultaneously scientific and social construction* of risk: modern science is one of the causes of risk, but it also defines and solves risk (Beck 1992, p. 155). According to Beck, we are increasingly concerned with the management of modern risks rather than with the transformation of nature itself:

> We are therefore concerned no longer exclusively with making nature useful, or with releasing mankind from traditional constraints, but also and essentially with problems resulting from techno-economic development itself. Modernization is becoming reflexive; it is becoming its own theme. (Beck 1992, p. 19)

Technological development, meant as a solution, becomes itself a problem. This is an interesting observation if we consider again the discussion about transhumanism: it suggests that we might work hard to release mankind from constraints, but this struggle against vulnerability—like all technological measures against vulnerability—creates its own problems, including new risks and vulnerabilities. I will return to this problem (or paradox) in the next chapters. Here I wish to use Beck to clarify what *social construction of risk* means today. The point is that risk and coping with risk depend on societal processes. Consider the introduction of new technologies. New technologies go hand in hand with 'the promise of security' which 'must be reaffirmed over and over again to an alert and critical public through cosmetic or real interventions in the techno-economic development' (Beck 1992, p. 20). This is especially the case for risks that concern the possibility of invisible harm: they are 'particularly open to social definition and construction' (p. 23). More generally, knowledge of risks always requires a socially mediated process of risk *production*; in this sense, there are no *bare* facts about risk. What we consider as risk and how we deal with it (including which risk issues we prioritise) depends on claims made by scientists and others in the public domain. There are 'definitional struggles over the scale, degree and urgency of risks' (p. 46), producing new risk antagonisms and risk communities.

Similar to the *psychologisation* of risk, this social constructivist approach first makes risk subjective but then objectifies this subjectivity. At first sight, it renders risk subjective, since it says that risk depends on the social, that is, on us. But then this social process is objectified: the sociologist can describe in an objective way the risk discourses, the risk societies, the contestation of risk in the public domain, the different kind of rationalities, and so on. What we feel about the risk does not matter, for it is again objectified as an element within social discourses, structures, and processes. We are insignificant actors on the societal stage of late modernity, which prestructures our risks and how we respond to these risks. The script has been written.

Moreover, even making risk subjective in this way seems to presuppose again an objective, real risk: there is *first* an objective risk which *then* is transformed by social processes and in social contexts. Although we may not have direct access to the socially and morally neutral *real risk* (our attention is drawn to the social-scientific identification, prioritisation, and reduction of risk; we cannot know risk outside this social context, and we have to remain within the social field within which risks are claimed and contested), it is assumed that there is still the risk-in-itself. Thus, a strict distinction is made between the risk-in-itself (objective, part of the natural world) and the *social construction* of risk (subjective, part of the cultural world).

The cultural-anthropological approach to risk seems to involve the same objectivist tendencies. Consider the work of Mary Douglas, who pioneered what is now known as the cultural theory of risk. In *Purity and Danger* (1966), she studied what is regarded as unclean or polluted (versus pure and sacred) in different societies and times, emphasising that the meaning of such words cannot be understood outside their social context. With Durkheim, she agreed that the psychology of the individual is not enough to understand society, including social controversies about risk. Culture has authority:

> Culture, in the sense of the public, standardised values of a community, mediates the experience of individuals. It provides in advance some basic categories, a positive pattern in which ideas and values are tidily ordered. And above all, it has authority, since each is induced to assent because of the assent of others. (Douglas 1966, pp. 39–40)

For risk, this view implies that what is considered dangerous and risky is defined by society and promotes certain social structures and institutions. For example, with Wildavsky, Douglas has explained political conflicts over air pollution and nuclear power in the USA in terms of a struggle between collectivist and individualistic ways of life (Douglas and Wildavsky 1982). At stake is not what can be called *natural* risk or the management of risk as such but the way we want to live our lives. More generally, Douglas has criticised rational choice and psychometric approaches to risk, which ignore how culture determines which risks are seen as worthy to take and try to depoliticise and de-socialise risk.

For example, in *Risk Acceptability According to the Social Science* (1985), Douglas complains that work on risk perception tends to see risk as an individual rather than a social phenomenon. In order to remedy this neglect of society and culture, she further develops *Purity and Danger*. She argues that 'social construction and consensus greatly influence human perceptions' (Douglas 1985, p. 38) and that social structure is a *moral system* that 'produces the different ways of categorising the physical world' and establishes 'the conceptual categories for public discourse' on risk (p. 39). Hence the perception of natural dangers is mediated by these moral-cultural categories. She shows that this does not only happen in *primitive* cultures—for example, when an earthquake is seen as a punishment for sins—but also in modern culture. The community determines which risks are grave or trivial, and institutions 'use the risk issue to (…) reinforce norms' (p. 92). Risk perception and decision-making about risk are at once cultural and moral:

> The question of acceptable standards of risk is part of the question of acceptable standards of living and acceptable standards of morality and decency, and there is no way of talking seriously about the risk aspect while evading the task of analysing the cultural system in which the other standards are formed. (Douglas 1985, p. 82)

In *Risk and Blame* (1992), Douglas continues this line of argument. She tells how, in the late 1950s, people who were enthusiastic about technology were asked to recognise real dangers and their causes by means of science, which made *real blaming* possible (Douglas 1992, p. 7). But when technology came under attack, the ideological and political dimension of risk was recognised. Our views of risk changed as society changed. This is also true for approaches to the study of our society, which change in symbiosis with society. For instance, cultural anthropology returned from exotic places and turned to its own (Western) culture, and Douglas has been part of that turn. Douglas argues that here too the evaluation of risk is 'a political, aesthetic, and moral matter' (p. 31). This does not mean that the dangers are not real—they are *real enough*, but we have to choose between them—and 'it would usually be preferable to have the choices directly presented as political questions, instead of sanitized and disguised in probability theory terms' (p. 39). According to Douglas, risk has been depoliticised and individualised (p. 40), but risks are always political and cultural: the question to ask is not how safe is safe enough for *me* but 'how safe is safe enough for this particular culture' (p. 41). She therefore approves of Beck's project to 'rewrite the whole of political science and economics in terms of risk' (p. 45).

While a cultural approach to risk helpfully criticises and complements overly individualistic and psychologist conceptions, rightly links risk perception to morality and politics, and is critical of objectivist conceptions of *the physical world*, it remains objectivist: not because it considers risk as objective or because it studies individual risk perception (it does neither) but since it comes in the form of a social *science*. The social-cultural mediation of risk perception is studied in an objectivist way. The issue of risk is put *into the context of the larger system*, which the anthropologist or sociologist claims to be able to oversee.

In taking this objectivist stance, Douglas follows Durkheim, who in his opposition to individual-psychological explanations of social phenomena defined social science as the science of social facts. Social facts are considered as things (Durkheim 1895). The sociologist is to reveal the laws of social structure. What is missing in this scientific approach to the social is the existential experience of risk and vulnerability, which cannot be captured in objectivist terms but requires a hermeneutical exercise. Let me start to work on this project in the next section.

3.2 An Existential-Phenomenological Alternative: A Relational Anthropology of Vulnerability

How can we benefit from the insights that (1) risk and vulnerability somehow belong to the human, that (2) our experience of risk and vulnerability matters, and that (3) risk and vulnerability are relative to the social and natural environment, without ending up with a view that risk and vulnerability are either entirely subjective or entirely objective, in the latter case perhaps supposing that there is a risk-in-itself? The following existential-phenomenological account of vulnerability is my attempt to meet this challenge.

3.2.1 Existential Vulnerability: Preliminary Phenomenology of Risk and Vulnerability

Let me first provide a summary of my claims. I start with the thesis that risk and vulnerability are neither subjective nor objective; instead, these terms tell us something about the *relation* between subject and object in the following way. First, the concept of *being-at-risk* is meant to communicate that risk is neither a feature of the world (an objective, *external* state of affairs) nor something that we create or perceive (a subjective construction by the mind, an *internal* matter), but is constituted in the subject-object relation. The same can be said of vulnerability. Second, the object can be real or imagined. What matters is that, being-vulnerable, we are always directed towards the object (the intentionality of being-at-risk). We have relations with the world, with others, and with the self. These relations render us vulnerable. It is our openness to the world that puts us at risk. It is our engagement with the world that renders us vulnerable. Third, we do not have full control over these objects and these relations and hence over our vulnerability. Yet we are not entirely passive with regard to these objects and these relations. These relations are transformed by us, that is, by humans as world-building (using technology), social (living in society), and outstanding (engaging in reflection) beings.

In a quasi-Sartrean style (but not adopting a Sartrean view—see later in this chapter), I add the following claims. First, we cannot escape the vulnerability of our existence by making risk subjective or objective. This means we have to face vulnerability *Angst* as opposed to fear for particular risks: we have to realise that we are existentially vulnerable and that we are naked. Therefore, it is a kind of *bad faith* to locate risk exclusively in the outside world (including the body we *have*—to use Merleau-Ponty's expression) or in our own mind. We have to face our being-vulnerable. Second, since we co-shape our vulnerability, we must also take responsibility for the design of vulnerability. However, this responsibility (and freedom) is not absolute, since we cannot entirely control the world, others, ourselves, or our relations to these. It is more about being *responsive* to the world and to others than about being a responsible, autonomous agent. I conclude that it is a task of ethics to understand and to help us shape our vulnerable existence, that is, our relation to the world, to others, and to our self.

Let me know these points step by step in the form of a preliminary phenomenology of vulnerability in order to achieve a better understanding of existential vulnerability.

Humans are vulnerable beings. The Latin word *vulnus* means *wound*. We can be hurt, and we can be wounded. The *capability* to be hurt is *vulnerability*; its *possibility* can be called *risk:* the possibility of being violated. I will use the term vulnerability to refer to the capability and possibility of being violated and wounded.

Using the term *vulnerability* emphasises the subject (its capability to be wounded), whereas using the term *risk* emphasises the object (and the possibility that it wounds us). I propose to use both terms in order to capture these different dimensions. On the one hand, our experience of vulnerability is always intentional in the sense that it is directed towards an object. Therefore, vulnerability is always also about the object: this is the *objective* dimension of vulnerability. On the other hand, although the term risk emphasises this object, we have no unmediated access to the object: the subject experiences the object in a particular way. Therefore, risk also has a *subjective* dimension. Thus, being-vulnerable and being-at-risk are both subjective and objective, or better, it is neither entirely subjective nor entirely objective. We are at risk.

Vulnerability and risk emerge from or within a relation between subject and object, and in this being-vulnerable and being-at-risk, the strict separation between subject and object made by dualistic modern thinking disappears. The possibility that I will be violated is neither entirely *in me* nor entirely *in the world*. The possibility arises from my standing in relation to the world, in my openness to the world, and in my engagement with the world—an engagement which precedes the modern construction of myself as an agent or a patient and a master or a slave.

We can be harmed emotionally or bodily and physically or psychologically. Words can hurt. Thoughts can hurt, too. Words and thoughts can be violent. Risk is not limited to technological or natural risk. There are many sources of risk and vulnerability (see below). Strictly speaking, one should not distinguish between these sources in terms of different objects but in terms of relations. This also implies that one can have different relations to the same object.

Furthermore, the term *engagement* is important in the description given above and needs to be understood in an active way. Vulnerability is not merely passive. To understand vulnerability as something entirely passive would be to turn the human being into an object once again or a *property* of that object. But *openness* does not mean passivity, and vulnerability is not merely a characteristic of our body or our mind. We are not vulnerable in the way a building or a bridge *is* vulnerable. Rather, we *make* ourselves vulnerable; we put ourselves at risk, by our mental and physical actions. We eat, we travel, we work, we love, we hope, and these actions make us vulnerable. Vulnerability, therefore, is not a property of the human person but a feature of the relation between us and the world. It is a feature of our way of being (in the world) and a way of existing. This is why I propose to use the term *existential vulnerability*. *Existence* is not a state but a doing; we should not use the noun but the verb: *to exist*. Similarly, *vulnerability* is not a state but something much more dynamic. The term *being* in *being vulnerable* or *vulnerable being* should be understood as a verb, as something ongoing rather than a state.

The ultimate source of vulnerability concerns the possibility of death. To live is indeed very risky: our chance of surviving life is zero. (This insight, that death is certain, creates additional vulnerability. See my remarks on imagination and on death below and in my introduction. And the fear of death is one of the key transhumanist concerns: they want to extend our lifespan or even make us immortal.)

Vulnerability is not a way of being exclusive to humans. Animals can also be hurt. Other living entities can be hurt, too, and even nonliving things can be called *vulnerable* if we broaden the definition of vulnerability to include the possibility of being damaged and violated. As living beings, we share vulnerability with (non-human) animals. (This is an important insight that has ethical implications for the treatment of animals.) But the additional vulnerability that arises from the specific way of existing and living as humans renders our vulnerability *human*. Our lives are not entirely different from those of animals or even plants, but they are partly and significantly different. The difference lies not so much in specific properties, or our essence, but rather in our way of being-in-the-world, our existence. We do not just act, but we are aware of what we do and (try to) give meaning to, or see meaning in, these actions. Moreover, humans have the ability to become aware of their own existence (they can become aware *that* they are), and therefore they can become aware of their vulnerability. This awareness does not only concern the past and the present, as well as the future. Thus, humans may reflect on the possibility of their own (future) death but also on other risks such as the possibility of being rejected by the ones they love or the possibility of not being able to realise their plans and their dreams.

Human beings are able to consider these possibilities since they have the capacity of imagination. Imagination enables us, among other things, to anticipate the actualisation of the possibilities of being violated. We are able to imagine that we get hurt. This awareness creates a kind of hyper-vulnerability or *second-order vulnerability*: apart from being-vulnerable in relation to the things we do and what might happen in a particular situation, we are also vulnerable by virtue of our capacity to *imagine* that something bad might happen. Our experience of existential vulnerability and existential risk does not exist separately from this imaginative act. Without imagination, I cannot become aware of my vulnerability. And second-order vulnerability really is a kind of *vulnerability*: there is always the possibility to be hurt in and through acts of imagination. The threat may (partly) constitute the actual harm, in which case vulnerability collapses into violence. This need not be physical violence. As I already noted, actions that make us vulnerable can be mental or physical. Words or images can contain the possibility of violence—whatever else they may do to us.

Vulnerability is not a property of the human individual (or, for that matter, any *separate* entity, any *individuum*) but, as I said, concerns a relation: something or somebody can potentially violate me. Let me further develop this point with regard to the emotional experience of vulnerability. In our experience of vulnerability, our consciousness is directed towards that which might hurt us and that which creates the possibility of violence. This object becomes the object of our fear. Since we tend to see vulnerability and risk not as a way of existing that arises from what we do but as something that threatens us from *the outside*,[1] our experience of vulnerability is characterised by fear *of something*. In the vocabulary of phenomenology, this is called intentionality. Our fear is directed towards the world. Fear is connected with hope. We fear that something will happen to us, and we hope that it will not happen. Hope and fear form a symmetrical pair.

Like imagination, emotions create hyper-vulnerability. Risk and vulnerability emotions are rooted in the second-order intentionality that happens as a result of the use of our imagination: we are directed towards an object, but this object is imagined and emotionally mediated, not directly related to the world. Our emotional experience of the imagined event itself is already able to hurt us, and thus an additional vulnerability arises. This role of emotions further explains *why* imagination

[1] To the extent that risk appears as something *external* (indeed as a *something* at all), the experience of vulnerability denies its very relational ground.

can hurt. By itself, imagination may seem a harmless mental activity. But imagination of risk does not happen without emotion. There is no *imagination in itself*, as a kind of neutral and accurate representation of the risk object. Imagination is embodied. Imagination involves emotion and vice versa. Our experience of vulnerability and risk is already mediated by our imaginative and emotional life. Before and even *at* the moment when vulnerability becomes violence and when the potential and *possibility* of being hurt are actualised—when risk becomes disaster and when possibility becomes event—our anticipatory imagination and our previous emotions co-shape our experience. For example, at the moment when I feel that a car accident starts to happen, I may realise that *it* will happen and is happening, where *it* is already preconstructed in the imagination, for example, when I imagine that the other car is going to hit me. Thrillers and (the better) horror stories play with this anticipatory imagination and how it affects our emotional experience of risk. As Hitchcock said, 'there is no terror in the bang, only in the anticipation of it.' Indeed, this role of imagination implies that for our *experience* of vulnerability, it does not matter whether the threat is *real*. The appearance of a threat, the imagined possibility of violence, is sufficient. Terrorism, for instance, can only exist because of this imaginative-emotional mechanism. For terrorists, it suffices that the public *feels* vulnerable to their (threat of) violence, which implies that a few violent acts may suffice for them to reach their aim. The threat is enough. The terror is not only in the actual acts but also in the influence these acts have on people's imagination, which creates second-order vulnerability: we know and we feel that we are vulnerable. (I will return to the topic of terror below, in my discussion of Heidegger.)

Furthermore, since existential vulnerability is a *relation*, there is no *objective* existential risk or *real* existential vulnerability if that presupposes a construction of existential risk or vulnerability apart from the human subject. We may want to construct *objective* risks and *real* vulnerabilities and use them for risk management purposes, but in so far as we are interested in what such measures and figures mean *for our existence*, we have to involve the subjective, imaginative, and emotional dimensions of human being. This implies that we can talk about risk management and a corresponding culture of risk but that we cannot meaningfully talk about *existential* risk *management* (unless perhaps we are prepared to include the idea of an economy of feeling and imagination in our conception of a household, which I guess would be an economy of excess rather than scarcity). Existential vulnerability and risk concern what might *happen* to us, which as an existential experience means realising the possibility of the happening of the unmanageable and the possibility of the end of control. In this existential approach, vulnerability and risk suggest the very opposite of *management* and the possibility that we can no longer manage the situation and no longer master ourselves and our lives.

However, existence is not about experience alone, if that means that we are passive with regard to that what can hurt us. As I said, the relation between ourselves and the world is shaped by what we do. We are not vulnerable in the same way as Sartre's cauliflower is vulnerable. We are never entirely passive. Vulnerability, then, is not only experienced but also created by doing the things we do. What we do has consequences for the existence and vulnerability of ourselves and others. From this realisation arises an ethics of care (or an ethics of security): we want to make sure that what we do does not hurt others (and does not hurt ourselves). We want to avoid bad things to happen *by taking care*. However, it is inadequate to restrict such an ethics to personal relationships, as is often suggested in so-called *ethics of care*. Vulnerability does not only arise from our relation to others (i.e. other humans) but also from our relation to other entities and beings, to collectives, and to *nature*—it concerns all our relations. Therefore, it is impossible to provide a comprehensive list of all sources of human vulnerability, since that would mean a list of all possible human actions and interactions, all human practices, and all human situations. The following, however, are some categories of risk and vulnerability, which are at the same time fields of care. Moreover, although I present them here as a kind of *objects*, they should always be understood in relational terms, for example, our relation to the body, our relation to nature, and our relation to others.

The Body. There is always the possibility that we become ill or die. (I will say more about this below.) What matters with regard to existential vulnerability is the relation we have to our body and to the possibility that something bad might happen to it—including the possibility that *the worst* might happen.

Emotions/the Passions. Our emotional life itself can often be the *object* of fear, that is, the way we experience and relate to our emotions is often a struggle and may hurt us. Not only do we have to cope with particularly *nasty* emotions connected with fear and suffering, such as hate and jealousy; the phenomenon of emotions itself—whether or not the emotion is *good* or *bad*—constitutes a problem for us. We feel that our emotions are not fully under our control. They *make* us say and do things we would not say or do in another state of mind. In this sense, our emotions (even the *good* ones) constitute a risk and render us vulnerable. This is why some philosophers have warned us of the danger of *the passions*. To moderns, it seems that emotions get in the way of rational thinking.[2] To the ancients, it sometimes appeared that a daemon took possession of them. Love can also make us vulnerable in this sense, if it releases irrational, unmasterable forces in ourselves.

Personal Relationships. Love and friendship render us vulnerable, since we fear the possibility that something happens to our beloved and to our friends and we fear the possibility of the end of the relationship. Once there is a relationship, something is at stake. Love and friendship make us vulnerable since we open ourselves to the other. This increases the possibility of violence (first-order vulnerability) and the possibility that we realise this possibility and suffer from this realisation (second-order vulnerability).

Values. As Martha Nussbaum has argued, ethics itself makes us vulnerable. When we value something, we fear loss of what we value. Valuing is a risky thing to do. But we cannot help valuing, and therefore we are always at risk in this sense. If nihilism were possible, we would *enjoy* moral invulnerability. But it is impossible;

[2] This is an inadequate and impoverished view of the relation between emotions and rationality. It is only possible to think of emotions in this way if we preconstruct them as radically separate from thinking.

we are condemned to valuing. Of course some philosophical schools, for example, the Stoics, have recommended disconnecting from *external* things. They were spreading the first seeds of nihilism.[3] But true nihilism is impossible; we cannot really disconnect. As long as we live, we have to value. We have to relate since we are relational beings. This means we can always be wounded (again).

Personal, Moral, and Social Identity. The relation to ourselves, in particular to what we consider to be our *identity*, is also a source of vulnerability. When I construct a certain identity for myself, which may appear to me as expressing what I *really* am, as my *authentic* self, I may get hurt when others say things that violate my self-image. At some times, the gap between our self-image and the image the other has of us may be unbearable and painful.[4] Generally, as sentimentalists like Adam Smith and Jean-Jacques Rousseau knew, since we care about what others think about us, in other words, since we are social beings, we can be hurt by what they say about us. Technologies can also change and threaten our self-image. For example, one issue in the discussion about human enhancement is whether or not such technologies would touch our personal identity. Žižek thinks genetic enhancement technologies would not touch our personal identities if we assume we are not determined by our genome (Žižek 1999). But things and technologies always influence what we think about ourselves and what others think about us.

Others, Society, and Politics. Other people, even if they are not directly related to us, can hurt us either emotionally or physically. We are vulnerable to various forms of *social* violence. People we know intimately can hurt us. Social structures such as family, peer groups, and organisations support us and contribute to our growth and flourishing, but they can also hurt us. There are group processes that go beyond personal interactions that can literally or figuratively kill us.[5] Political systems and institutions can hurt individuals, either directly (e.g. by all kinds of forms of repression) or indirectly, by their policies (e.g. a policy that fails to tackle extreme poverty) or lack thereof.

Religion. Religious organisations can hurt us, for instance directly by punishing people or by isolating individuals from the rest of the community or (indirectly) by advocating behaviour that hurts (others), for example, those who do not belong to the group. But religion can do more harm, perhaps harm of a *deeper* kind. A powerful religious imagination may increase our vulnerability. Religion can help us to cope with vulnerability and risk. It can offer consolation, and it can offer techniques to deal with physical or mental pain. But it can also hurt and create additional suffering. In religions with a personal god, people are subject to the risks arising from their personal relation with that god, who may punish or reward them. If it is a love relationship they have with the god, they are vulnerable to the joys and risks of love. Love creates dependence; love can hurt. The same is true for polytheism: with many

[3] Recently, this connection between Stoicism and nihilism has also been recognised in Dreyfus's and Kelly's book *All Things Shining*: the authors write that 'Roman Stoicism is grandfather to the nihilism of the secular age' (Dreyfus and Kelly 2011, p. 65).

[4] In his novel *Nachtzug nach Lissabon*, Pascal Mercier reflects on this experience.

[5] See, for example, Girard's work on scapegoat rituals (Girard 1972 and subsequent books).

gods, one never knows what all of the gods are up to. There is always the possibility of violence. Furthermore, there is also a religious transformation of the fear of nature. The god(s) incorporate(s) the forces of nature (which are not experienced as such but as divine), which may *bless* but also destroy.

Nature/the Environment. What we call *nature* is an important source of risk. As far as we experience our own bodies as *nature*, this can include risk of sickness and death or being bodily, physically hurt (see also being hurt by other people). And there are many risks from *outside* that threaten us (*outside* meaning not part of the *human community*), ranging from hurricanes and tsunamis that destroy entire cities to wild animals and to a bad harvest due to drought. However, like the other risks and vulnerabilities, *natural* vulnerability is not merely something *external* that has nothing to do with how we live. Rather, it belongs to how we actively relate to the world and to what we do in the world. The particular face of risk depends, for instance, on whether and how we engage in hunting, agriculture (including how we deal with animals), or mining or on where we (decided to or are forced to) live, for instance near the water, in a region with a high risk of floods. And as I mentioned before, *natural* risk is often entangled with *technological* risk, as in the case of climate change and other environmental hazards. *Environmental* risk and vulnerability depend on how we relate to our environment, actively and passively. Vulnerability is also a kind of *doing*, with both a *passive dimension* (something might *happen* to us) and an *active* dimension: something might happen to us because we live in a particular way, because we *do* particular things, and so on.

Moreover, *natural* vulnerability is often entangled with *social* and *political* vulnerability, and *place* matters with regard to these vulnerabilities. This is especially clear in the case of disasters. *Natural* and *social* are so interconnected that it is difficult to speak of disasters that are only *natural* or *environmental*. People with different social backgrounds live in difference places, and hence their vulnerability to disasters differs as well. Moreover, vulnerability to disaster also depends on one's political environment: there might be failures in emergency response by the state or failures in social support systems. Consider the hurricane Katrina disaster in 2005, which flooded 80 % of New Orleans and caused many deaths. Cutter writes about the disaster that there was 'a lack of political will to redress social inequalities' and that it manifested 'a confluence of natural and social vulnerabilities', due to geography: 'the spatial interaction of humans and their environments over time' (Cutter 2006). Indeed, there are studies about the geography of social vulnerability that show how vulnerability to environmental hazards is socially and geographically mediated. As Cutter et al. put it, 'vulnerability to environmental hazards means the potential for loss. Since losses vary geographically, over time, and among different social groups, vulnerability also varies over time and space' (Cutter et al. 2003, p. 243). Vulnerabilities to disaster are partly socially geographically created. Hence, the possibility that a so-called *natural* disaster happens is never a merely *natural* risk; it always has social and geographical dimensions. Think also about the earthquake in Haiti in 2010: the impact of an earthquake depends on who lives where and on how societies are organised. Poor people are often more vulnerable to floods, for example, because they live in places where there is a high risk of floods.

They also have houses that are not built as well. And they may lack the financial resources to adapt to hazards. As Beck already argued (Beck 1992), we should attend to the (social) distribution of risks (see also Chap. 3); learning from the *geography of vulnerability*, we should add that *place* matters here. Your vulnerability depends on how you live but also on how much money you have and on *where* you live. Consider also big cities. Uitto has summarised the problem with megacities as follows:

> It is becoming obvious that megacities are highly vulnerable to all kinds of environmental disasters and that vulnerability is increasing. (...) Megacities contain very large concentrations of people, buildings, industry and finance. They have become so complex that they are almost impossible to manage. Moreover, the political and economic centrality of megacities ensures that the consequences of the disasters that occur will be very far-flung. (...) It has also been pointed out that it is increasingly difficult to identify urban disasters that are purely either 'natural' or 'technological' or 'industrial' or 'social' or 'biological'. (Uitto 1998, pp. 8–9)

Consider also the climate change problem: how people might be affected depends partly on their social background and on *where they live*. In geography, there have been studies about 'how and why places and people are vulnerable to environmental hazards' (Cutter 1996, p. 530). (And one can do the same for non-human vulnerabilities, for example, with regard to the risk of loss of species and their habitats. Abbitt et al. use the term *hot spots of vulnerability* for areas vulnerable for future species loss (Abbitt et al. 2000). The geography of species affects their vulnerability.) Thus, so-called *natural* or *environmental* vulnerability has significant social and geographical dimensions.

Economy and Finance. We are dependent upon economic, financial, and globalised systems that engender their own risks. In capitalist market economies, there is always the risk of an economic and financial crisis. We might no longer feel that we are in the hands of the gods but that we are at the mercy of *the markets* and other visible and invisible hands. Again, it is important to understand that the risks are related to what we do, including what kind of systems we create. This is not to say that *chance* and *fate* do not play a role in economy and finance; rather, like other tragedies, the tragedy of economic risk and financial vulnerability arises from this combination of what we do and what happens beyond our control. Particularly, tragic is that we come to experience systems that we created, such as technological and economic systems, as autonomous forces that are entirely *outside* ourselves and our control, similar to *nature*. But there is always a mixture of activity and passivity in the tragic.[6] In the next chapter, I will argue that our technological, economic, and political actions and systems have created new vulnerabilities or, rather, that they have transformed our bodily and psychological vulnerabilities. In fact, human responses to vulnerability have always transformed our vulnerabilities. In the next chapters, I will elaborate this point and discuss its implications for an ethics of the posthuman future.

---

[6] See, for example, my article on Kierkegaard, tragedy, and technology (Coeckelbergh 2011).

Making Sense of the World. Vulnerability can also be *metaphysical* and *hermeneutical* in the following sense. To think about the world and to interpret the world is to put oneself at risk, and we suffer if we fail to make sense of the world. If and to the extent that we search for a meaningful order and if we feel that we do not find this in our (contemporary) world, we may feel alone and vulnerable in what we come to experience as a cold, scientific universe. We may also feel lost in our *inner* universe, which may be full of dangerous abysses and dark depths or which may reveal itself as fragmented and confused. (Both kinds of vulnerabilities are typically *modern* and arise from the assumptions that the world is disenchanted and that there is such a thing as an *inner* realm.) And even if some of us find some metaphysical security in religion or philosophy, there is a risk of losing that security. Belief may disappear. Thinking is also risky. If, as Sloterdijk thinks, truth is an immune system, there is no guarantee that the system will remain working. Finally, can we make sense of the technologies we created? Human enhancement technologies can be seen as an attack on our ethical thinking and on our efforts to make sense of ourselves. If we cannot make sense of these technologies and of the posthuman future proposed by transhumanists, we suffer. Note that vulnerability can be experienced and *done* by individuals but also by collectives. We can call a group of people, an organisation, or a nation vulnerable. There too, the vulnerability depends on what the collective *does*. The object of vulnerability is at the same time a subject, or, rather, vulnerability has an intersubjective dimension. My vulnerability is deeply related to your vulnerability because the same or similar things might happen to us and because we do similar things. (However, to the extent that we do things differently, our vulnerability differs as well. I will discuss the issue of cultural differences in the next chapter.)

In a way similar to other work inspired by the existential-phenomenological tradition, the concept of existential vulnerability questions, and attempts to move beyond, several important modern distinctions and dualisms.

First, the concept of vulnerability crosses the categories of body and mind. The experience of the possibility of suffering cannot be detached from physical and material aspects in two ways. First, suffering itself, when actualised, will have a physical or material component—even if there is only a *mental* or *psychological* violation the brain and other parts in the body are involved in the experience. Second, the anticipation of suffering will have, to some extent, an emotional aspect, and emotions involve, again at least to some extent, a *bodily* change.[7] Moreover, *even* our body is not something we have *direct* access to; rather, we experience our body through cultural, political, and technological media. Even *physical* pain is not unmediated but interpreted by us as the persons we are, living in this society and this culture, etc. (which does not necessarily render it any less *real* or painful). Thus, there is no mind without a body and no body without a mind. Being-vulnerable is *in the mind* and *in the body* at the same time. Moreover, as I already indicated, being-vulnerable is neither internal nor external: it is about my *relation* to the world.

---

[7] In contemporary philosophy of emotion, there is a discussion between cognitive and bodily change approaches to emotions. I take a *compatibilist* position here.

Note also that the analysis of sources of existential vulnerability cuts through nature-culture distinctions. Since vulnerability is at the same time subjective and objective, and internal and external, it is as much about people as it is about things and as much about culture as about nature. Latour is right that politics is not about humans alone but also about things (Latour 1993). A politics of vulnerability, therefore, must concern itself with various kinds of entities, human and non-human (see Chap. 7). Similarly, vulnerability is about various kinds of entities to which we relate. However, this relation is not one of a *subject* that perceives and is influenced by an *object*: in the process of engagement with the world, the object is no longer a *pure* object and the subject is no longer a *pure* subject. As said, being-vulnerable crosses the subject-object distinction. There is no object separate from our subjective risk experience; at the same time, there is no subject separate from the risk it is subjected *to*. The *vulnerability subject* is constituted by the *vulnerability object* and vice versa. The concept of existential vulnerability developed here allows for such a non-dualistic, relational perspective, since it is defined in such a way that it does not make sense to talk about a subject without mentioning the object.

There is a contextual, situational aspect to risk and vulnerability. For example, the bad harvest mentioned above is not only related to the modalities of agriculture and culture (what we do and how we interpret what we do and give meaning to it) but also has different consequences in different social and political arrangements. The same is true for financial crises: the way we cope with them differs, and the very definition of a *crisis* and the very idea *that* there is a crisis depend on context. The related vulnerabilities, therefore, will differ accordingly. (For example, it seems to me that the very idea that there could be a *crisis* itself generates a kind of second-order vulnerability.)

This analysis is nonobjectivist but also avoids subjectivism and hence hopes to contribute to the ongoing development of the philosophical tradition of phenomenology in the following way. The concept of intentionality was meant to bridge the gap between subject and object, but *in practice*—that is, if we consider the writings of most well-known phenomenologists—phenomenology did not entirely succeed in doing this, since it tended (tends) to focus too much of the subject. When it does so, it is too idealist or even solipsist. The concept of existential vulnerability conceptualises a stronger link between subject and object. But does it go beyond the distinction? What kind of relationality is this?

One interpretation is that vulnerability emerges out of the tension between subject and object. Only in the actual event of violence, perhaps, the distinction between object and subject disappears: there is no longer a tension when the space between object and subject collapses. When woundability becomes being-wounded, it might happen that the subject becomes objectified and that the object is swallowed by subjective experience. But vulnerability is mainly defined in terms of the *possibility* of violence, not its actualisation.

Another interpretation is that the concept of existential vulnerability already goes beyond the subject-object distinction in the experience of the possibility of violence, since, as I wrote before in this experience, the subject cannot be defined without the object and vice versa. I think this more *radical* interpretation must be preferred, and I introduced the term being-at-risk to support this direction of thought (see also my use of Heidegger in the next section).

Finally, is this a *pessimistic* view of human being? Not necessarily. Although in accordance with common usage of the words risk and vulnerability I have defined existential vulnerability in terms of potential *harm* and *suffering*, we should not forget that it is the same vulnerability that enables us to experience pleasure, beauty, and love. Our vulnerability gives us not only the possibility of violence but also the possibility of good. There is always the possibility that good emerges. Stronger: *good could not emerge without vulnerability*. Existential vulnerability is a precondition for human good and human excellence.

Similar points have been made in the literature. For example, McIntyre has highlighted in *Rational Dependent Animals* (1999) how much we need others, and we should not see this dependency (and corresponding vulnerability) as a defect but as a good. Nussbaum has made a similar point about the relation between ethics and vulnerability in *The Fragility of Goodness* (Nussbaum 1986). I will return to this issue in Chap. 5 and in my discussion of one of Houellebecq's novels in Chap. 6.

At this point in the argument, some may want to take a *shortcut* to ethics: why not take human vulnerability as the foundation for ethics? But what does this mean? The mere *fact* of human vulnerability cannot be a *foundation* for ethics if that is to mean *providing a reason to act and think ethically at all*. If there is such a reason or such a foundation at all (which I doubt), I doubt if it can be provided by philosophical reflection.

One could argue that *if* we want or *decide* to act and think ethically, in other words *if* we already take the ethical point of view, then we *must* do something about vulnerability. We can *decide* to care. However, there is also a sense in which we cannot help caring, valuing, etc., as I said before. There is an ethics that emphasises the will, but there is also an ethics that grows out of our existential condition. Experiencing vulnerability as vulnerable beings gives rise to care: care about what might happen to us and to others, which may or may not result in caring *for* ourselves and for others, in giving care.[8]

Ethics is a response to human vulnerability, a way of coping with it. But it is not the only way; humans have different ways to cope with vulnerability. I doubt if *any*

[8] The relation between vulnerability, caring about, and caring for is of course a matter of discussion. In this context, it might be interesting to consider Levinas's view that the vulnerable, naked face of the other makes an ethical demand on us (Levinas 1961). In *the total nudity of his defenceless eyes*, Levinas reads the ethical demand not to kill (Levinas 1961, p. 199). Whereas modern moral philosophy argues about reasons for doing something for others and about ethical principles, it seems that in Levinas's view there is a direct relation between human vulnerability and ethics, *before* all reasoning and discourse. Whereas modern ethics assumes that logos is prior to ethics, Levinas seems to turn this around. First, there is responsibility, understood as a response to the face of the other; then there is discourse and decision, which tends to turn the other into an object and the situation into an ethical *case*. In this sense, perhaps, vulnerability can be seen as a foundation of ethics: the vulnerability of the other asks me to respond. However, when Levinas then argues that we gain access to infinity by opening one's self to the other, he seems to turn to abstraction again (the other becomes the Other) and to the word (logos) and its religious source.

response that tries to reduce vulnerability is ethically good by definition or if *reducing* vulnerability is always ethically good at all. Let me postpone a full discussion of the relation between vulnerability and ethics. In the next chapter, I will discuss some ways of coping with vulnerability and assess if they really manage to reduce human vulnerability. In Chap. 5, I will discuss how we should view the relation between vulnerability and ethics.

3.2.2 Existential Vulnerability: Being-at-Risk, Fear, and Care (Using Heidegger 1)

In the previous section, I claimed that vulnerability is a feature of our being in the world and I proposed an *existential* approach. In order to further develop what this means, let me reformulate and elaborate my alternative epistemology of risk and vulnerability by using Heidegger's philosophical language. In *Being and Time* (1927), Heidegger used the term *being-in-the-world* to argue that we should not split up subject and object. We are not Cartesian egos that transcend the world; we are always already in the world; and we are *engaged* in the world. Moreover, we are also always already social:

> The clarification of being-in-the-world showed that a mere subject without a world "is" not initially and is also never given. And, thus, an isolated I without the others is in the end just as far from being given initially. (Heidegger 1927, p. 109)

Being-in-the-world is always also a being-with-others. In terms of knowledge, this view implies that knowledge is not about representing an objective world but has to do with what Dreyfus calls *the embodied coping* (Dreyfus 2006): we are always already involved in the world.

If we apply this Heideggerian phenomenological approach to risk and vulnerability, then risk is not something that is *out there* in the world and vulnerability is not an objective feature that can be known separated from our subjectivity. Rather risk and vulnerability are constituted in and by our engagement with the world and with others. Furthermore, the *I* that is at risk and that is vulnerable is not given *before* that engagement with the world. We are at risk and are vulnerable as we are in the world and as we engage with the world. Being-in-the-world involves being-at-risk. As we cope with our environment, we put ourselves at risk and we render ourselves vulnerable. Risk and vulnerability belong to our kind of being—with *being* understood as a verb and as a relation. Since our being is always worldly and social, we are vulnerable.

With Heidegger, we can also say more about vulnerability and death. I already said that human beings can reflect on the possibility of their own death and that renders them vulnerable. The imagination of death constitutes a kind of second-order vulnerability: even if we are not immediately threatened by death, it remains the *ultimate risk* since we know that we live towards death. Heidegger links death to our being always ahead of ourselves. We face possibilities. Death is the ultimate possibility. But this possibility is not *external* in the way an event is external; it belongs to what Heidegger calls *Dasein*. Heidegger uses the term being towards death. As I mentioned in my introduction, becoming aware of this involves *Angst*: 'In *Angst*, Da-sein finds itself faced with the nothingness of the possible impossibility of its existence' (Heidegger 1927, p. 245). With this term, Heidegger does not mean the emotion of fear but the disclosure of 'the most extreme possibility', which is 'the ownmost nonrelational (…) possibility of Da-sein' (p. 239). Thus, the ultimate danger is non-relationality, which is nothing else than the end of your being: death. Without relationality, there is no vulnerability and there is no human being.

Further interpretation of the latter claim (especially ownmost) would require me to say more about the individualist-existentialist side of Heidegger's work. There are alternatives, for example, Levinas sees the vulnerability and death of the *other* as primordial. However, as I will explain in the next section, I will not incorporate this dimension of Heidegger in my discussion of risk and vulnerability. For the phenomenology of risk and vulnerability as developed here, I propose to look closer at two other themes: fear and care.

Heidegger analyses the phenomenon of fear in a way that helpfully contributes to the phenomenology of vulnerability and risk I wish to articulate. How do we experience risk? First he says that *the fearsome* is the harmful, which is coming near: 'As something threatening, what is harmful is not yet near enough to be dealt with, but it is coming near. As it approaches, harmfulness radiates and thus has the character of threatening'. Threatening means that *it can get us and yet perhaps not.* It may pass by. But this 'does not lessen or extinguish fearing, but enhances it' (Heidegger 1927, p. 132). Thus, if we fear a risk, we fear it because it presents itself as a (real) *possibility*. It may happen and it may not happen. Furthermore, when what threatens is unfamiliar, fear becomes horror. And when it is suddenly there and fear becomes alarm, it becomes terror (p. 133). In other words, we come to experience risk and vulnerability as fear, as horror, or as terror—depending how *familiar* we are with the danger and how *near* the hazard is. Furthermore, fear is not something extraordinary but is part of our human existence. All modifications of fear 'point to the fact that Da-sein as being-in-the-world is "fearful"' (pp. 133–134). This is how we are attuned to the world. I conclude from Heidegger's analysis that risk experienced as fear belongs to our existential structure. It is not just a specific risk and a specific fear; these kinds of experiences are part of our being-in-the-world.

We can also use Heidegger's language to say more about the relation between risk and care. According to Heidegger, being-in-the-world is a matter of care; different ways of being-in-the-world have the character of care: 'to have to do with something, to produce, order and take care of something, to use something, to give something up and let it get lost, to undertake, to accomplish, to find out, to ask about, to observe, to speak about, to determine' (p. 53). To care means both *to take care of* something or someone and to be *concerned*. All these ways of being-in—understood as ways of caring—also create risk and vulnerability. Care is a response to risk, but as *caring* (the feeling and the activity), it also creates risk. If we were not involved in anything, then there would be no risk. If we were not concerned about anything or anyone at all, if we did not care *about*, we would not be vulnerable. But we *are thrown in the world*, and we are with things and with people.

We are involved; we are engaged in the world. We do things and we care. Therefore, risk and vulnerability belong to our existential condition as beings that take care and care about.

3.2.3 Existential Versus Existentialist (Not Using Heidegger 2)

Before I continue my argument concerning existential vulnerability, let me further clarify the existential approach I take. In order to think about risk and vulnerability, we must be able to describe it. But a major advantage of an existential approach to vulnerability is that it allows us to avoid scientism: our struggle with our vulnerability *can* be described, but not in terms of *facts*, calculations, and numbers. In order to do justice to the phenomenology of risk and human vulnerability, a more literary, narrative approach seems to be better in place.

For example, in *Existential Anthropology* (2005), Jackson—as an anthropologist—writes about events that demonstrate the precariousness of human existence. His examples show that while every one of us must try to live our lives, we often experience failure. Indeed, we could say that the possibility of failure is also a kind of vulnerability. There is always the possibility that things work out differently and that we cannot realise our goals. But he does not describe this in the terms of objective science but in a more narrative way. As he explains in *Things as they are* (1996), his approach is phenomenological, by which he means that he attempts to describe consciousness in its *lived immediacy*. One of his themes is freedom. Influenced by Sartre, Arendt, and others, he sees human being as a relationship between 'forces that act upon us and our capacity for bringing the new into being' (Jackson 2005, p. xi), involving 'endless experimentation in how the given world can be lived *decisively*, on one's own terms' (Jackson 2005, p. xii). Another (related) theme is the social. Interestingly, he claims that our ambiguous relation to non-human nature and technology—our *fear* of it—stems from our social fears:

> We fear the inert, unresponsive, silent and alien appearance of the extrahuman world because we see mirrored and magnified in it our deepest *social* anxieties – that we will not be recognised, that our voices will not be heard, that our actions are without significance, that we are mere means to genetic ends that far outreach the time-scale of our conscious lives, and that we possess no more meaning than grains of sand on a beach or flotsam along a tideline. (Jackson 2005, p. 124)

Thus, as an anthropologist, Jackson considers the problem as less conceptual (the blurring of the distinction between nature and culture and between the familiar and the alien) and more an inherent condition of the intersubjective, interpersonal, *social* life. Our fears of technology are linked to our fears of the strange, the alien, the new, and the other. He understands technology in the framework of our existential struggle to find a balance between controlling our fate and accepting the limits of our control. He explores how we actually experience and interact with technologies (Jackson 2005, p. 127), again with a stress on the link with the intersubjective. For example, he observes that computers (that cause difficulties) are addressed by their users as if they have a reciprocal relationship with a person and people will use emotional strategies that change their relationship with the computer in the same way as they would do with people (Jackson 2005, p. 133). He also discusses how people involved in organ transplantation have difficulties starting and tolerating a relationship with the not-self, with the radically other, and how they employ different strategies to nevertheless make the relationship viable (Jackson 2005, p. 136).[9]

Jackson is interested in social alienation. His (hypo)thesis is that 'the ways in which we experience our relationships with both persons and machines will depend upon *the degree to which we feel in control of these relationships, as well as the degree to which these relationships are felt to augment rather than diminish our own sense of wellbeing*' (Jackson 2005, p. 131). Not everyone agrees with his first criterion (control). In Chap. 5, we will see that ethics of care theorists question an emphasis on autonomy and control, although perhaps some of them may agree with Jackson's weaker criterion that we find 'some sense of balance between being actors and being acted upon' (Jackson 2005, p. 182). Furthermore, the emphasis on autonomy is distinctively modern, which runs contrary to the approach to existential vulnerability I take in this book. In addition, it seems to me that Jackson overemphasises the social dimension of technology. But his existential-phenomenological approach to the issue is a welcome change, and much of what he does (and the way he does it) *is* in line with my approach in this book. Moreover, he poses an interesting question with regard to the discussion about Western modernity: 'To what extent can we *live* with boundary blurring?' (Jackson 2005, p. 139). The answer must be, I think, that anthropologists know that many people on earth can do that very well, whereas 'Western', modern people have a problem with blurring such boundaries.[10]

In any case, when it comes to his view on technology, Jackson's work can *explain* (but not necessarily justify) some of the opposition against human enhancement technology: such technology can be experienced as an obstacle to our desire for meaning, control, and categorisation since it may give us the feeling that *technology takes over*. (At the same time, the project of human enhancement can be seen as the summit of the desire for control and the ultimate attempt of modernity to control everything—even human nature.) Moreover, Jackson's focus on personal narratives of coping with our existence is an approach that can inform a descriptive philosophical anthropology of vulnerability. Perhaps we need fewer risk *reports* and risk *assessments* and more narratives that help us cope with risk and vulnerability.

However, an existential approach to vulnerability need not imply an *existentialist* view. There are, of course, some methodological similarities. I used the existential-phenomenological turn to overcome a merely naturalistic account of vulnerability. Therefore, I agree with Sartre and Heidegger that we must shift the anthropological focus from *essence* to existence. To use Heidegger's words (Heidegger 1927),

---

[9] Note that he criticises Ihde for not exploring this social dimension.

[10] Note that this distinction between nonmodern and modern has been questioned by Latour in his *We Have Never Been Modern*: as the title says, he argues that we have never been modern in the first place.

I could agree that *Dasein* has no *nature* but that its own being is an issue for *Dasein*. I have also articulated some aspects of being-vulnerable by using Heideggerian language.

Nevertheless, while I learn from the existential approach, I hesitate to align my approach too closely to Sartre and Heidegger, since I do not subscribe to the further normative project of *existentialism* and their full philosophical-anthropological and ethical views. In this sense, I use *Heidegger 1* (Heidegger's existential-phenomenological approach) but not *Heidegger 2* (Heidegger the existentialist, including his views on freedom and authenticity). Consider Sartre's and Heidegger's views of freedom. For Sartre, the shift to existence implies that I have to *choose* myself. This choice and responsibility are absolute, and the burden we face as a result is heavy. Heidegger, by contrast, can be interpreted as saying that existence is not a matter of choice at all and not a matter of practical deliberation, finding reasons for action, beliefs, etc., either, but rather of coping as being-in-the-world. In this non-individualistic interpretation of *Being and Time* (informed by Heidegger's later work), *Dasein* is not the Sartrean absolute and isolated subject but is part of a background of practices and an opening in which being lights up. Actually, *Dasein* is no longer a subject at all. And whereas Sartre seems to be on the side of those who want to struggle against vulnerability, Kierkegaard asks us to accept vulnerability. Furthermore, Sartre, Heidegger, and Kierkegaard all aim at authenticity.

In my analysis of vulnerability, I am sometimes inspired by Sartre and Heidegger, but I do not fully adopt their views on freedom and human being, do not make a claim about authenticity, and postpone normative judgment about what kind of attitude we should take towards human vulnerability. First, the analysis above was meant to avoid the view of the human as an entirely active, isolated, willing, individual subject but also its opposite: an entirely passive entity determined by whatever happens to it. Vulnerability has to do with what happens to us and with what we do, with the inner world and with the outer world, with trying to take distance, and with being part of the world. We are both active and passive in relation to vulnerability, and it is the *relation* between subject and object that is presupposed by existential vulnerability (although this understanding of vulnerability renders the gap between the two smaller than moderns like it to be—it even tends to blur it). Moreover, the struggle against vulnerability is not only an individual matter but also a collective project. Second, although phenomenology has the ambition to focus on how the world appears to us, its *theories* are often very abstract, far removed from concrete, daily human experience. Third, I do not make a claim about what would be the most *authentic* way to cope with human vulnerability. This is not only because that concept is highly problematic (where do we find this authenticity, in our *inner* being?), but mainly since when I analyse the concepts of risk and vulnerability (and, in the next chapter, transformations of vulnerability), I do not want to *prescribe* a certain way of coping with vulnerability but to *understand* (our coping with and evaluation of) vulnerability and its relation to technology. Just as I do not argue for or against human enhancement, I do not argue for or against vulnerability or for or against a particular attitude towards vulnerability. Rather, instead of evaluating vulnerability in general, I will argue in the next chapters (Part II) that we must evaluate specific vulnerability transformations and I discuss how we can do that. If we have a responsibility and a freedom as beings-at-risk, it is the responsibility and freedom to shape our vulnerability and to cope with our vulnerability—at least to the extent that we can. To use a Sartrean term, it may be indeed a kind of *bad faith* not to accept this responsibility and to locate vulnerability entirely in the object or in the subject. However, this responsibility and freedom cannot be absolute, since there is much that I (as an individual) and we (as collective) cannot control. There are limits to the malleability of human vulnerability. There are limits to the *design* of the human and of human vulnerability. Nevertheless, as beings that are aware of our vulnerability, we are *forced* to take a stance towards our individual, collective, and human vulnerability. We must not try to escape that vulnerability *Angst*. This is a *prescription*, perhaps. It is certainly a normative stance. But it is not an argument for or against vulnerability, for or against human enhancement, or for or against specific technologies. This book is meant to contribute to a better *understanding* of the problem of human vulnerability and of the possibilities of an ethics of vulnerability that takes into account the influence new and emerging technologies (will) have on our lives. This can indirectly inform decisions concerning practical coping with risk and vulnerability, but this enormous existential task is a task for each of us individually, for our institutions, communities and collectives, and for future generations. Philosophical reflection can contribute to that task by clarifying what is at stake and by opening up different ways of thinking about it.

3.2.4 The Tradition of Philosophical Anthropology: Plessner

Let me end this chapter with a note on how my approach relates to the tradition of philosophical anthropology. As I take Helmuth Plessner's work to be most relevant to my project, let me briefly comment on his main theory and concepts.

My account of vulnerability is largely compatible with Plessner's anthropology developed in *Die Stufen des Organischen und der Mensch* (1928), although he does not discuss existential vulnerability. I agree with Plessner's objection to Heidegger (Plessner 1928, p. 22), with Plessner's existential approach, and with his rejection of naturalist explanations of the origin of culture (e.g. Plessner 1928, pp. 392–393). Let me take the liberty to use his concepts to further explain and clarify my own position:

Eccentricity: according to Plessner, humans do not stand in their centre but find themselves outside themselves (e.g. Plessner 1928, p. 364). They can take distance from themselves, since they are already standing outside themselves. They stand in nothingness; they have a *utopian* residence (Plessner 1928, p. 424). Using and developing Plessner, we could now interpret this basic existential structure as a precondition for vulnerability. If we were not standing outside ourselves, we could not imagine ourselves as the one to whom something might happen. It is because we stand outside ourselves that we can experience ourselves as vulnerable. The experience of existential vulnerability presupposes our standing out. Being outstanding, we know that we are at risk.

Plessner also mentions fear and, in particular, the fear of death (Plessner 1928, p. 393). According to him, these fears are symptoms of the basic structure or mode of eccentric positionality. In contrast to animals, humans live in the future (Plessner 1928, p. 394). Fear and care already presuppose the human life form, characterised by eccentric positionality. We may agree with this. However, Plessner's account omits a role for the imagination. Without imagination, we would not be the kind of beings that care about, or fear, the future. The capacity of imagination is a condition of possibility of eccentric positionality and of the experience of existential vulnerability.

Aussenwelt, Innenwelt, and Mitwelt. When Plessner distinguishes three spheres of existence, I propose to interpret these as possible *sources* or *worlds* of vulnerability: vulnerability can emerge in our relation to the outside world (*Aussenwelt*), our relation to ourselves (*Innenwelt*), and our relation to others (*Mitwelt*). However, these *worlds* should be understood as relations, not as separate spheres. The concept of existential vulnerability, then, firmly links the three worlds. For instance, our fear of what others think of us makes us vulnerable, and this fear depends on others as physical beings present in *the outside world*, depends on our own image of ourselves (*inner world*), and is about our relation to others. And in the very experience of vulnerability, distinctions between these *worlds* disappear: we are vulnerable in one world. To say it in a Heideggerian way, we are vulnerable as we *world*. To *world* is to render oneself vulnerable.

Furthermore, my approach differs from Plessner's since his emphasis is more on stability than on change and more on *nature* than on *culture*. Although he takes an existential approach, he is still mostly interested in what humans *are*; he sees them as *naturally* incomplete (Plessner 1928, p. 396) and talks about *natural artificiality* (Plessner 1928, p. 383). I agree with these views (see also the previous chapter), but I doubt that talking about *nature* and *natural* is the best way to articulate the existential approach. Plessner succeeds at rejecting a naturalistic approach, but at crucial points, he still uses the language of essentialist *human nature* approaches to express his position (next to the language of *being*). Talk about nature is misleading if it suggests too much stability and if it neglects existential experience (by suggesting an objectivist understanding of the human). Given these problems, I think it is better to make a complete shift to the language of *being*, without thereby neglecting concrete, human experience (as Heidegger did in his abstract human philosophy of *Dasein* and *Being*).

I found in Plessner's book only one passage that vaguely refers to existential vulnerability (and to a famous biblical passage, which is about a very specific kind of vulnerability): since humans have knowledge, they see that they are naked, and therefore they lose their directness and instinctive certainty and need to live via artefacts (Plessner 1928, p. 384). Thus, in his view, mediation by technology is the result of (forbidden?) knowledge. Plessner seems to suggest that culture and technology, indeed technological culture, are a punishment by God. Does this mean he thinks that there is the possibility of direct knowledge and unmediated living—that we can even *think of and conceptualise* this possibility? This would contradict his own view about the *natural artificiality* of humans, which suggests that unmediated access to reality and unmediated life is not possible. Moreover, the myth of the lost directness assumes that it would be better for us if we could get rid of technology altogether. But if technology makes us human, then a *direct* way of living would be a non-human form of life. Nevertheless, we can learn from Plessner that the experience of vulnerability (being *naked*) presupposes that we are not one with the world and that in this sense we are always already alienated and that we experience the world as eccentric beings. I can only see that I am naked if I take distance from myself, if I realise that I *have* a body, and if I am not in my centre. However, at the same time, this eccentric position and this alienating experience of vulnerability do not alter our basic existential condition of being-in-the-world, which is a precondition for vulnerability.

Finally, Plessner thinks that we do not know what we are doing unless we experience it by means of history (Plessner 1928, p. 418), that is, afterwards. This is an interesting suggestion for my reflections about an ethics of the technological future: If we do not yet know the precise consequences technology will have in the future, is such an ethics *possible* at all? I will discuss this problem in Chap. 6.

References

Abbitt, Robbyn J.F., J.Michael Scott, and David S. Wilcove. 2000. The geography of vulnerability: Incorporating species geography and human development patterns into conservation planning. *Biological Conservation* 96(2): 169–175.

Beck, Ulrich. 1992. *Risk society: Towards a new modernity*. Trans. M. Ritter. London: Sage Publications.

Coeckelbergh, Mark. 2008. Risk and public imagination: Mediated risk perception as imaginative moral judgment. In *The ethics of technological risk*, ed. L. Asveld and S. Roeser. London/ Sterling: Earthscan Publishers.

Coeckelbergh, Mark. 2011. Moral responsibility, technology, and experiences of the tragic: From Kierkegaard to offshore engineering. *Science and Engineering Ethics* 18(1): 35–48.

Cutter, Susan L. 1996. Vulnerability to environmental hazards. *Progress in Human Geography* 20(4): 529–539.

Cutter, Susan. 2006. The geography of social vulnerability. Published on-line at http://understanding-katrina.ssrc.org/Cutter/.

Cutter, Susan L., Bryan J. Boruff, and W. Lynn Shirley. 2003. Social vulnerability to environmental hazards. *Social Science Quarterly* 84(2): 242–261.

Douglas, Mary. 1966. *Purity and danger: An analysis of concepts of pollution and taboo*. London: Routledge, 2003.

Douglas, Mary. 1985. *Risk acceptability according to the social sciences*. New York: Russell Sage.

Douglas, Mary. 1992. *Risk and blame: Essays in cultural theory*. London: Routledge.

Douglas, Mary, and Aaron B. Wildavsky. 1982. *Risk and culture: An essay on the selection of technical and environmental dangers*. Berkeley: University of California Press.

Dreyfus, Hubert L. 2006. Overcoming the myth of the mental. *Topoi* 25: 43–49.

Dreyfus, Hubert, and Sean Dorance Kelly. 2011. *All things shining. Reading the Western classics to find meaning in a secular age*. New York: Free Press.

Durkheim, Émile. 1895. *The rules of sociological method*. Trans. W.D. Halls. New York: The Free Press, 1982.

Girard, René. 1972. *La Violence et le sacré*. Paris: Bernard Grasset. Trans. P. Gregory. *Violence and the sacred*. Baltimore: The Johns Hopkins University Press, 1979.

Heidegger, Martin. 1927. *Being and time: A translation of Sein und Zeit*. Trans. Joan Stambaugh. Albany: State University of New York Press, 1996.

Heidegger, Martin. 1977. The Question Concerning Technology. In: Heidegger, Martin. The Question Concerning Technology and Other Essays, trans. W. Lovitt. New York: Harper & Row.

Jackson, Michael (ed.). 1996. *Things as they are. New directions in phenomenological anthropology.* Bloomington: Indiana University Press.

Jackson, Michael. 2005. *Existential anthropology: Events, exigencies, and events.* New York/ Oxford: Berghahn.

Latour, Bruno. 1993. *We have never been modern.* Trans. C. Porter. Cambridge, MA: Harvard University Press.

Levinas, Emmanuel. 1961. *Totality and infinity.* Trans. A. Lingis. Dordrecht: Kluwer Academic Publishers, 1991.

MacIntyre, Alasdair. 1999. *Dependent rational animals.* Chicago: Open Court.

Nussbaum, Martha. 1986. *The fragility of goodness: Luck and ethics in Greek tragedy and philosophy.* Cambridge: Cambridge University Press, 2001.

Plessner, Helmuth. 1928. *Die Stufen des Organischen und der Mensch: Einleitung in die philosophische Anthropologie* (Gesammelte Schriften IV). Frankfurt am Main: Suhrkamp, 1981.

Slovic, Paul. 2000. *The perception of risk.* London/ Sterling: Earthscan Publications.

Uitto, Juha I. 1998. The geography of disaster vulnerability in megacities. *Applied Geography* 18(1): 7–16.

Žižek, Slavoi. 1999. Die Nacht der Welt. *Suddeutsche Zeitung.* 23 September 1999.

Chapter 4
Cultures and Transformations of Vulnerability

4.1 Culture(s) of Vulnerability

4.1.1 Experience: Imaginations of Vulnerability

The attitude we take towards vulnerability depends on our representations and imaginations of it. In this section, the question is not only 'how do I imagine myself being hurt by this or that particular object', or 'how do I imagine the possibility of that particular event', which is the imaginative dimension of being vulnerable which I explained in the previous chapter, but also what kind of attitude we take towards existential vulnerability in general. The conscious experience of vulnerability is not a single happening, but occurs frequently at many different times. As a result, we discern patterns of vulnerability. Moreover, we are the kind of beings to whom things do not just happen: we take an attitude towards what happens and try to *give* meaning to those experiences and *find* meaning in those experiences. These imaginative-representational, cognitive, meaning-giving, and meaning-finding processes can be called *cultures* of vulnerability. I write 'cultures' rather than 'culture', since these processes vary with time, place, and the people involved. Cultures of vulnerability, then, are shaped by attempts to imagine, understand, and give meaning to human vulnerability. This can take place at various levels: persons, groups, organisations, professions, societies, and 'civilisations'. Going beyond particular experiences of vulnerability, a culture of vulnerability reaches a certain level of abstraction and distance and generates ideas—for instance, ideas about death. Representations, too, start leading their own lives in the form of art. But at the same time, these ideas and representations do not stand alone from experiences of vulnerability; they colour subsequent experiences, forming a feedback loop. Cultures of vulnerability—at various levels—co-constitute our vulnerability experiences. We *live* in cultures of vulnerability. I already suggested above that our imaginative and emotional vulnerability experience is mediated by our 'mental' life. But this mental life does not stand alone, separate from the culture in which we live. Our imaginative and

M. Coeckelbergh, *Human Being @ Risk: Enhancement, Technology, and the Evaluation of Vulnerability Transformations*, Philosophy of Engineering and Technology 12,
DOI 10.1007/978-94-007-6025-7_4,

emotional experience of vulnerability is culturally mediated. For instance, not only our personal history but also the history of our society is relevant to vulnerability experiences. In addition, cultures shape our experiences as well as our *practices*. In the previous chapter, I argued that vulnerability cannot be understood apart from action. What we do *makes* us vulnerable. But action is not just about single actions. Just as there are patterns in our experience of vulnerability, there are also patterns in our actions, usually called habits. And actions and habits are part of a constellation of actions, objects, and ideas. Let us call them practices. Practices of vulnerability, then, are intertwined with experiences of vulnerability.

First, I will further explain the imaginative-representational dimension of vulnerability cultures, and then I will say more about vulnerability *habitus* and vulnerability *praxis*.

4.1.1.1 An Example: Experiences and Cultures of Health and Illness

Let me illustrate and elaborate on my analysis of vulnerability by discussing vulnerability related to health and illness.

Vulnerability related to health and illness means the possibility of becoming ill, the actualisation of which involves both physical and psychological suffering. I experience my vulnerability by imagining that possibility. This imagination of vulnerability is connected to (other) personal experiences. We have been ill before, perhaps we have seen someone who suffered from the same illness, or someone we knew died after a long illness. Imagination is coupled with emotion, in particular fear. We fear becoming ill; we fear death. We hope to stay healthy; we hope to live.

The possibility of death seems to be always on the horizon of our vulnerability experiences. Or, rather, death is the end of all horizons, the end of my world. Our imagination does not stop at the possibility of illness, and the experience of the possibility of illness will be unconsciously coloured by the shadow of the possibility of death. Against Epicurus,[1] we must locate the fear of death—and in a sense, the suffering *from* 'death'—in its anticipation: it is always its *possibility* we fear; death itself (i.e. being death) knows no suffering. In death there is no violence; the violence and the harm lie in the imaginations of suffering and in the suffering before death.

Vulnerability to illness and vulnerability to death only exist since we are active beings, since we live. Being ill or dying is only a possibility if we are living and if there is an alternative to illness and death: health and life. Illness is the loss of health, and death is the end of life. Both pairs must be discussed together. To say something about illness is to say something about health. When we are healthy, we tend to forget about both health and illness. Often we only think about health risks and health vulnerability if some harm has already been done.

Illness can be experienced in at least two ways, depending on how we relate to our own body. As Merleau-Ponty (1945) argued, we simultaneously have a body

---

[1] Epicurus famously argued in his *Letter to Menoeceus* that since death is deprivation of sensation, there is nothing terrible in death and it should therefore not concern us.

(the body is an object for us) and are a body (our body is part of what we are; it is a lived body). Similarly, we can experience 'having an illness' and 'being ill'. We can experience illness as something we have and something we received, something from outside, an intruder in our body. But we can also experience being ill in a more 'internal' sense: we live our illness, and the illness is part of what we are; in this sense we are our illness. Both are varieties of vulnerability experience.

Our experience of illness is shaped and mediated by culture. At individual level, there are imaginations of illness, discernments of patterns, and meanings we give. We also have ways to deal with vulnerability to illness, perhaps performing a health or wellness ritual, practice sports, etc. These ways of coping are connected to macro-cultures and macro-practices of vulnerability. For instance, in our societies, representations of healthy young people are part of a culture of youth and health, which involves certain practices and mediates our experience of illness. Individual experience of illness is co-shaped by this culture. For example, if old age illnesses such as Alzheimer are represented as terrible (e.g. as a black hole in which human self-consciousness and subjectivity disappear and are forever lost), then such a representation will influence decisions people make about the end of their lives. If the value of individual choice is emphasised in our culture, if vulnerability practices are medicalised,[2] etc., this has consequences for the way we experience and deal with our vulnerability.

Representations of human vulnerability objectify and materialise vulnerability; they alienate from concrete experience and practice, for example, in philosophy or (medical) science books (objectification), but also in art (materialisation). Consider the *vanitas* theme and the generous attention for illness and death in the history of art and in contemporary art. One of the most controversial recent developments is the project of the German artist Georg Schneider, who raised controversy in 2008 when he said he wanted to exhibit a dying person (or a person who had just died) in an art gallery or museum in order to show the beauty of death.[3] Many people think death does not belong in such a context. Death is often seen as 'the last taboo' (The Times, April 23, 2008), which tells us much about how we cope with our vulnerability: one way of dealing with vulnerability to death is creating a culture of youth and health and repressing the perception of death. At the same time, we see that *violent* death is overrepresented on the screen of our violent culture. Death is exhibited and covered up at the same time. In this paradoxical culture of death and vulnerability, we live.

Can we avoid disease and perhaps death? Transhumanists explicitly understand their movement as a struggle against disease and death. As I said in Chap. 2, in his 'Fable of the Dragon-Tyrant' (Bostrom 2005), Nick Bostrom seeks to support the case for human enhancement by representing disease and death as a dragon-tyrant that demands human sacrifices. If this metaphor is appropriate, then of course we must all become transhumanists. But is it? Opponents of transhumanism may represent illness and death in another way, for instance as parts of normal life, not as something external

---

[2] Medicalisation of a practice means that problems in that practice come to be defined as medical problems.

[3] See, for example, (http://www.telegraph.co.uk/culture/art/3672813/Artist-Gregor-Schneider-plans-to-put-dying-person-on-show.html)

to it, and as aspects of life we can give meaning to. Each side has its own metaphors and moral repertoires, which are part of different vulnerability cultures.

For instance, a typical metaphor that is used in discussions about new technology (including human enhancement technologies) is that of a 'monster', an image that is often related to the Frankenstein narrative. If we call a technology a monster, it means at least three things: (1) we strongly and morally disapprove of it, (2) we see it as something that is external to our own activities, and (3) we see it as not fitting in our culture as a whole. In this way, technology is often overly dissociated from our lives and our culture, which prevents an adequate analysis of technology and its influence on our lives. Opponents of transhumanism may discuss posthumans in terms of 'monsters' in order to express their view that such human enhancement produces beings that are radically different from what we 'naturally' are, that it produces something that is entirely alien to what we know, and that such a transgression of limits is unacceptable. By representing disease and death as a monster (a tyrant dragon), Bostrom uses a similar dissociative strategy, constructing disease and death as something alien to the human condition. Whatever side we choose (if we have to choose between these options at all), the point I want to make here is that our individual and collective vulnerability experiences (and practices) are not to be understood as single and isolated events, but are culturally embedded and must be analysed as such.

4.1.2 Praxis and Habitus: Imagination as Representation Versus Imagination in Action

Cultures of vulnerability should not only be understood as imaginations of vulnerability in the sense of *representations* of vulnerability. Being-vulnerable and being-at-risk is a matter not only of experience and representation but also of action and habit. As I argued before, it is as much about what we do as it is about what we experience. Let me develop this view.

My previous arguments must be understood as a way to shift the concept of vulnerability from a passive, static, isolated possibility experience and event to a dynamic, long-term dimension of practice that draws on, and co-shapes, the culture we live in. Moreover, to understand coping with vulnerability as practice and habit also means that this coping is historicised: how we deal with vulnerability now depends on our past experiences of vulnerability and our past efforts to cope with vulnerability. Meaning is historically produced.

So far, however, vulnerability itself might be taken to be invariable. In the previous discussion, 'culture' is still too much a way of seeing, perceiving. Surely, I said that culture shapes our practices of vulnerability. And practice shapes culture. Experience and culture, or practice and culture, do not simply mirror one another but change one another. But, so one may ask, what about vulnerability 'itself'? Under the description provided above, it still appears that there are experiences *of* vulnerability and practices *of* vulnerability—in other words, it still appears that human vulnerability is stable whereas only the cultural representations, beliefs, meanings, experiences, and practices of that same, shared human vulnerability are variable in time and place. But if vulnerability is so much bound up with practice, how *could* it be stable? Rather, in so far as it is action/practice dependent and culturally shaped, it seems to move in time and in place. If vulnerability is so connected with the way we live and give meaning to our lives, it turns out to be as changeable as human life itself. It is *lived* vulnerability, and living changes and ways of living change. Our 'vulnerability habits' and 'vulnerability practices' surely have some degree of stability—this is inherent in their definition—but they can also change; they have a history. Here culture is less concerned with imagination as representation (e.g. representations of vulnerability) than imagination in action: our creative coping with vulnerability. There is no vulnerability in itself, and there is no vulnerability apart from culture as a living process.

In the next section, therefore, I take one more step beyond an essentialist understanding of vulnerability by introducing the concept of vulnerability *transformations*. Human culture is not only a reflection of vulnerability, but also shapes and changes it and is itself shaped and changed by it. There are not only varieties of vulnerability in place but also in time. Human culture can be seen as partly resulting from vulnerability transformations: the ways we cope, have coped, and will cope with vulnerability changes our vulnerability and co-constitutes what we call 'culture'—in particular a vulnerability culture. This concept of vulnerability transformations also better suits the nonessentialist concept of human being I drew on at the end of the first chapter. 'Existence' should not be objectified. Perhaps 'existing' is a better word: it is not something that is, not a 'something' at all, but a verb, a doing. And I will argue that part of what we do is transforming our vulnerability, thereby transforming ourselves, our culture, and our world.

Note that local and temporal difference does not imply that nothing remains stable. Cultures often share habits and practices (there is temporary stability), and there is a shared human vulnerability (culture) since we all have human bodies and hence share the same bodily vulnerabilities. As Shakespeare puts it in *The Merchant of Venice*, when he has Shylock argue for similarity between Christians and Jews, 'If you prick us, do we not bleed? If you tickle us, do we not laugh? If you poison us, do we not die?' (Act III, Scene I). There is a shared, cross-cultural experience of human vulnerability that spans a longer period—ranging from a generation to that period we call 'human history'—and is not restricted to particular places and societies. However, what we share is restricted neither to 'bodily' vulnerability as 'natural' vulnerability (as what I said before may suggest) nor to 'essential' features of what we are as humans. What remains stable—between different cultures and to a certain degree between different times—is not so much who or what we are 'by nature', but at some level of abstraction *how* we are and how we *relate* to our vulnerability, including how we relate to what we call our 'body'. All humans live vulnerable lives and partly share how they—as a person, a group, a society, a species—*struggle* against vulnerability. We partly share a vulnerability *habitus*.

The term 'habitus' is particularly helpful, given the existential-phenomenological orientation of my inquiry. To talk about the perception or experience of vulnerability may give the false impression that we are concerned with 'representations' or 'objective structures' of vulnerability; instead, vulnerability is always *lived vulnerability*. But this does not mean that vulnerability is merely 'subjective'. Bourdieu used the term 'habitus' to avoid having to describe the social world either in terms of 'objective' sociological laws or in terms of 'subjective' mental experience. Whereas objectivism 'constitutes the social world as a spectacle offered to an observer who takes up a "point of view"', Bourdieu focuses on the dispositions or habits that are 'constituted in practice':

> It is possible to step down from the sovereign viewpoint from which objectivist idealism orders the world (…). To do this, one has to situate oneself *within* (…) the practical relation to the world, the pre-occupied, active presence in the world through which the world imposes its presence, with its urgencies, its things to be done and said (…). (Bourdieu 1990, p. 52)

Thus, habitus is neither entirely 'external' nor entirely 'internal' but cannot be described in terms of these dualist categories. In the action and in the practice, the 'objective' and the 'subjective' merge. Habit is a kind of know-how, which is learned in the past and which gives a style to practices: it is a way of doing things. Usually we are not aware of it: habit is embodied and only questioned when a particular way of doing no longer works.

For vulnerability, this means that we should neither describe its 'objective' structures from the 'outside' nor reduce it to subjective experience. Instead, vulnerability habitus is a way of doing, a way of coping with vulnerability which we embody as members of a particular culture and as members of the human species. To put the human vulnerability habitus into words is to describe a way of doing: 'this is how we humans deal with vulnerability'. But since these ways of doing are embodied and usually not reflected upon, they are only partly open to description. Often we can only *show* how we deal with vulnerability. We can show the human struggle, for example, by means of art, literature, and film.

We also share some part of our vulnerability struggle and habitus with animals and other non-humans, but humans have specific ways of coping with human vulnerability and human cultures have specific ways of doing so. Some cultures stress the differences between humans and non-humans. For example, in Western culture there is a discourse on the 'essential' anthropological difference, which depends on distinguishing ourselves from non-human others: usually animals, in the past people with another skin colour, in the future perhaps humanoid robots. We define ourselves in terms of what we are not. But such a discourse says much about the way we struggle against the vulnerability of our conception of the human: we try to set up conceptual security structures, built to withstand the storms and floods that we fear will erode and finally erase the human figure from the beach of history (to adapt a metaphor from Foucault). This is how traditional-style humanism tries to defend the human: by means of a *via negativa*. It is a negative anthropology: it defines the human by saying what it is not—not an animal, not a robot, etc. For example, in his *Discours de la méthode* (1637), Descartes defined the human as *not* a beast-machine (*bête-machine*).

The discussions about transhumanism and robotics, for instance, can be seen in this light. If we come to understand that some animals are very 'close' to humans, perhaps we need new monsters and aliens in order to say what we are (not).

The imagery of posthuman life forms (see Chap. 6) or of cyborgs, among other imageries, may serve this aim. And of course there is cultural variety in how we live, but there is an underlying pattern: fear for the end of humanity is part of Western culture. It is part of our human identity to struggle in a particular way, to be *this* kind of vulnerability warriors rather than others, and modernity makes possible its own kind of vulnerability warriors. (And maybe animals and other living beings could also be defined not as members of an essentialist category, having such and such characteristics, but rather in terms of a mode of being, involving a certain way of struggle with vulnerability.) If our human way of struggling were to change radically, we would be different beings altogether. It remains an open question whether 'transhumanist' human enhancement would have that effect. In any case, vulnerability changes are changes to a mode of *being* (understood as a verb) that has not only a location but also a history and a future. Furthermore, it is a future we can influence to some extent. What we are as humans is neither a 'nature' that is fixed once and for all by a divine creation or intervention—beliefs which had (and have) a powerful influence on our struggle with vulnerability and change(d) who we are—nor a 'nature' that is studied and constructed by the natural sciences and from which we can comfortably distance ourselves, but *a way of being*, which for humans involve living, experiencing, and transforming that being. In contrast to what Aristotle thought, there is no pre-fixed *telos*. We cannot use a philosophical, theological, or scientific 'GPS' to arrive at a preset aim. What humans will become will emerge and grow. The future of humanity is neither completely malleable (as some transhumanists may think) nor completely fixed in advance. We use and need both moral-anthropological engineering and moral-anthropological imagination to explore better ways of being for us as humanity and for us as a particular society and community—with imagination understood as representation and as coping in practice. But we cannot make a blueprint which we then simply use to make 'products'. It is neither sufficient nor desirable to make a 'concept' of the future of humanity and the future of society, which we then implement. In this sense, the future of humanity is not a matter of *design*. Coping with vulnerability is always 'experimental'. We are part of the experiment. Subject and object cannot be clearly distinguished.

According to Plato, Heraclitus said that we could not step twice into the same river. Everything flows (panta rhei). But 'we' also flow. The human flows. We are the river.

4.2 Vulnerability Transformations

The concept of existential vulnerability (and existential risk) can be employed in an account of the origins and history of human culture, including magic, religion, science, and morality. Dewey wrote in *Experience and Nature*:

> The world is a scene of risk; it is uncertain, unstable, uncannily unstable. (…) Anthropologists have shown incontrovertibly the part played by the precarious aspect of the world in generating religion with its ceremonies, rites, cults, myths, magic; and it has shown the pervasive penetration of these affairs into morals, law, art, and industry. (Dewey 1929, pp. 41–42)

Modern institutions, however, must also be understood as responses to existential risk. According to Dewey, science is also a kind of 'magic':

> Our magical safeguard against the uncertain character of the world is to deny the existence of chance, to mumble universal and necessary law, the ubiquity of cause and effect, the uniformity of nature, universal progress, and the inherent rationality of the universe. These magical formulae borrow their potency from conditions that are not magical. Through science we have secured a degree of power of prediction and of control [...]. But when all is said and done, the fundamentally hazardous character of the world is not seriously modified, much less eliminated. (Dewey 1929, p. 44)

Dewey understood risk as a property of the world, rather than a feature of our (active) relation to the world, as I proposed. And his use of the word 'magical' suggests too much that our efforts to struggle against vulnerability have not really changed the world. Moreover, he describes 'magic' from an external point of view, as an uninvolved observer. Therefore, his view of risk and vulnerability is still too 'objectivist' and naturalist. But his insight that risk is not eliminated but rather *modified* in our efforts to secure a degree of control is also applicable to vulnerability. We try to diminish vulnerability, but only succeed in changing it. Our responses to vulnerability transform that vulnerability and transform our existential condition. However, our strategies to cope with vulnerability are not limited to religion and science, as Dewey's quote may suggest. Humans cope with their vulnerability in myriad ways, which correspond to a spectrum of interrelated existential vulnerability transformations. We have used religious, social, political, economic, technological, and psychological strategies in order to reduce vulnerability, but we have only modified it.[4] Once we realise that this is the case, we try to find a new balance and begin a new effort. The history of this vulnerability struggle is the history of what we call culture.

Without trying to be comprehensive, let me discuss four important anti-vulnerability strategies and explore how they transform and transformed our vulnerability: spiritual technologies, material technologies, social technologies, and technologies of the self. I call them 'strategies' and 'technologies' since they have an instrumental aspect (they are understood as a means to an end) and they have a material aspect (they involve material means and alter—among other things—material conditions). The strategies respond to four kinds of vulnerability relations: our relation to the supernatural, our relation to nature, our relation to others, and our relation to ourselves. In all cases, an order is created with the aim to avert the chaos, the contingent, and the uncannily uncertain.

4.2.1 Spiritual Technologies and Religious Culture

As Dewey noted, fearful existence has generated 'religion with its ceremonies, rites, cults, myths, magic'. Religion creates and maintains a divine order. It tries to tame

[4] Perhaps there are not only qualitative transformations; maybe they also have a quantitative direction. Is there a 'dialectic' of culture at work (in the sense Adorno and Horkheimer used the term), by which I mean that the very strategies to overcome vulnerability only increase it? Does the cumulative accretion of layers of vulnerability, or *extensions* of vulnerability, alter the human way of being and human existence?

natural violence by ritual and magical means. However, religion is a response not only to our relation with nature but also to our relation with others and ourselves. It is also meant to control social chaos (due to e.g. rivalry and jealousy) and—in modernity—the uncannily unstable within ourselves. By creating an additional relation, one between us and the supernatural, it transforms these vulnerabilities into religious vulnerability. Our struggle with nature (and with ourselves) is reshaped into a struggle with the gods and the spirits. Our struggle with others is transformed into the communal relation with (the) god(s). What might happen to us then concerns what the gods might do to us.

Religion also controls social violence. For example, it transforms the more violent versions of scapegoat rituals into what are perceived as less violent forms: the victim becomes a 'surrogate victim' (an animal or an object). The scapegoat ritual itself, too, can be seen as an anti-vulnerability strategy. We transfer or delegate our vulnerability to the scapegoat; the possibility of violence comes to rest completely on its shoulders. This restores social, communal harmony. This is my interpretation of the immunity mechanism described by René Girard in *Violence and the Sacred* (1972) and subsequent works. The scapegoat ritual is the 'metamorphosis of reciprocal violence into restraining violence through the agency of unanimity' (Girard 1972, p. 96). We seek immunity from violence by allowing a little homoeopathic violence, as a kind of social vaccination. In the ritual imagination, violence is seen as something exterior to humans, a virus intruding in our society. Indeed, Girard describes the scapegoat ritual in analogy with vaccination: 'The physician inoculates the patient with a minute amount of the disease, just as, in the course of the rites, the community is injected with a minute amount of violence, enabling it to ward off an attack of full-fledged violence' (Girard 1972, p. 289). For this purpose, the scapegoat 'candidate' must be at once an insider and an outsider, a good guy and a bad guy at the same time—comparable to the requirements for a tragic hero, who is supposed to be a little bit good but yet weak enough to have a 'tragic flaw' (Girard 1972, p. 291).[5]

Furthermore, our struggle with ourselves is changed into a relationship with the god(s). The supernatural is responsible for (keeping) order in nature, in society, and within ourselves. Rituals should confirm that order and please the gods. To use Girard's words again, 'For if we neglect to feed the god, he will waste away; or else, maddened by hunger, he will descend among men and lay claim to his nourishment with unexampled cruelty and ferocity' (Girard 1972, p. 266). Of course, there is more to religion and society than the scapegoat ritual or this nourishment imagery. But in all religions, ritual is of central importance with regard to social, psychological, economic, and spiritual order and harmony, and therefore to the attempted reduction of social, economic, psychological, and spiritual vulnerability.

In the ritual, the divine and the spiritual (the other world) are related to material objects and practices (this world). In some religions, both worlds are more closely

[5] Note, however, that Girard even sees all individual death as a way of keeping up the community: 'The death of the individual has something of the quality of a tribute levied for the continued existence of the collective. A human being dies, and the solidarity of the survivors is enhanced by his death' (Girard 1972, p. 255).

linked than in other. Sometimes objects or humans can travel between both worlds (or embody both). As I said before, there are many vulnerability cultures.

However, it seems that most of these efforts to reduce risk and vulnerability are in vain. Instead of more order, which was and is at least *one* (often inexplicit) aim of religion and ritual, a new kind of vulnerability is created. Humans now can be hurt by the supernatural, directly or indirectly—through humans. Their fragile bodies and minds become potential playing fields for the gods and for those who claim to mediate between humans and gods. Humans are hurt and wars are fought in the name of religion. In modernity, people struggle with and because of their personal relation to the one god. They feel very vulnerable: will they receive the god's blessing, his grace? Or will they be abandoned by the god? These people are vulnerable to anything that can happen in a human relationship. Religion can enhance the quality of (some) people's lives—as I said religion can provide consolation and vulnerability is also a condition for good experiences—but it can also degrade them, violate them, tear them apart. More generally, apart from harming individuals, religion can harm or even destroy the natural, social, and psychological order it was meant to maintain or enhance.

4.2.2 *Material Technologies and Technological, Financial, and Economic Culture*

In the development of material technologies as anti-vulnerability strategies, we can detect a similar pattern. Initially the purpose of these technologies (today called 'science and technology') is similar to that of magic in relation to nature: we want to get a grip on our relation to nature, and more: we want to increase our control, our mastery of nature, in order to reduce the vulnerability that arises in the relation between us and nature.[6] We try to get a cognitive and material grip on natural violence. We want to understand and control the natural order by means of science and technology. An artificial, technological order is created as a result of this; we want to have a safe, controlled environment in which to live.

Furthermore, although technology responds to vulnerability in the context of (what we moderns view as) our relation to nature, it has also implications for our relation with the supernatural, with others, and with ourselves. Since modern technology tries to make humans little gods, the relation with (the) god(s) changes. If we can create and maintain our own material order, is there still a need for the supernatural? Moreover, technologies change societies dramatically. The rapid development of medicine and mass production of consumer goods have created

[6] Note the not so common relational dimension of vulnerability in this formulation. Turner writes that technology and culture can be seen as mediations between the scarcity of natural resources and the vulnerable human body (Turner 2007). But this approach wrongly understands vulnerability as a feature of the body, rather than as a feature of the relation between us and the world.

large social entities and corresponding institutions (bureaucracies, states, armies, etc.) to control people and to control society. Finally, technologies reshape how we relate to ourselves. The material economy seems an external answer to our internal economies of desire, which make us vulnerable. The market economy diminishes our vulnerability; It seems that the market economy diminishes our vulnerability by giving us freedom - the freedom to choose what we want. The result is a technological, financial, and economical culture. But is it an order? Like religion, it has also created chaos and has modified—perhaps added new layers to—our vulnerability.

Let me start with the financial and economical 'order'. First, modern people feel that without the supernatural we are thrown upon ourselves: we feel lost in a spiritual desert, which leaves us vulnerable, perhaps more vulnerable than ever before. Market economies try to respond to such desires and create new gods and a new order, on which we become dependent ('the market', financial systems, consumer goods). In response to vulnerability, we have replaced the gods with insurances and financial investments. Second, the market society has replaced relations in small communities (the gift) by trade relations (contractual exchange, money), but this means we have become dependent on remote processes and on people we do not know. We have even become dependent on machines and software, on algorithms, to which trade actions are delegated. Third, the market economy has established order in our household of desire, but at the cost of us being vulnerable to whatever happens on the market. A crisis may always happen. We do not know when it will start, and if it happens, we do not know when it will end. If politicians, bankers, and insurers ask for more money from us, we are as helpless as the worshippers of a god who asks sacrifices. We keep on giving all we have. We can only pray and hope. We have to surf on the waves created by 'the market'. Who can remain standing? The globalisation of the economic and financial order has added to this vulnerability: our 'skin' is now extended all over the globe; our globalised body and mind is vulnerable to what happens thousands of miles away. Investment decisions are made in distant places. Money and goods circulate all over the place. There are global streams of information. Economic chaos and financial crisis somewhere else influence our lives here. Vulnerability processes were always holistic, but now we start to realise it.

Technological culture promises to give us more control. We invent machines to have (meta-)control. But instead, we now have the feeling we live in a 'risk society' (Beck 1992). We see 'risk' everywhere. We are afraid of the technological systems we created and on which we depend for our lifestyle. The contingency that we try to modify by means of religion has now moved down from the heavens to the earth: to the technological artefacts, systems, and infrastructures we use. Nuclear plants can explode, cars can crash, and computers can be infected. Technology itself, meant to tame fate, has become the locus of tragedy (De Mul 2006). It is the 'place' where we experience contingency and vulnerability. Fear of nature has first been replaced by fear of the gods and then by fear of technological catastrophes. Our societies have become highly dependent on technology. And our own minds and bodies, increasingly safe from biological viruses, are extended via the Internet and in this way become vulnerable to virtual viruses, system breakdowns, and other ICT 'security' problems. It seems that in response, our immune system must now be extended too: we need a hybrid biological and material immune system. Moreover, we relate to others and to ourselves through the medium of technology, which has transformed the vulnerability we experience in these relationships. Our avatars co-shape our identities, and we open up ourselves to people we have never met. Social networks render us socially vulnerable in a different way than before, but we remain vulnerable. The material and virtual order (which itself has a material basis: PCs, network cables, etc.) is as unstable as the natural order was.

Furthermore, we believed that we had mastered the natural order, but it turns out that the technological, social, financial, and economic systems we created are highly dependent on the natural resources and climate of the earth, which may not be able to sustain the lives of future generations if we continue our lifestyles. Our system of energy provision turns out to be more vulnerable than expected, and our system of food production turns out to be riskier than we believed. We have tried to create energy security and food security, but by doing so we have increased our environmental vulnerability. We are now threatened by climate change; we are highly dependent on global markets for our food, and we know that now the industrial processes of farming and food production are highly vulnerable. The possibility of disease and death is always there. In sum, we have the feeling that we are more dependent, uncertain, and vulnerable than ever before. We got rid of (most of) the wild animals that threatened our existence in the early days of human history, but we have created moving things that are at least as dangerous. There are new snakes and new lions waiting for us in the jungle of cyberspace. We have exiled the dark by electrical lights, but the dark is ready to return in our homes, our societies, and our minds. We fear the shadows of technology. We have tried to reduce our physical vulnerability by medical means, but new diseases have emerged, our technological environment has created new risks, and our longer lives are no less vulnerable than those of our ancestors. We have changed our existential condition, but we have failed to expel vulnerability. We are still human. And in our struggle with vulnerability and in the creation of new vulnerabilities, in the style of these attempts, in the style of these failures, and in the style of these new beginnings, we are very, very human.

Note that my view differs from that of Beck and Giddens in at least two important ways. First, I discuss existential risk, which is something that arises in the relation between me and the world, in the experience of human being, whereas as I argued in Chap. 3, Beck and Giddens seem to have a more 'objective' understanding of risks (according to them, risks can be 'external' or 'manufactured'). I argued that existential risk is never 'external', we experience and imagine it, and in our practices and in our habits, we are at risk as beings engaged in the world. (Note also that this approach does not necessarily deny, but at least does not *depend* on a distinction between the real and the imagined.) Second, Beck and Giddens reserve the term 'risk society' for a society like ours that is very much and increasingly occupied with risk. But I do not claim that our society is especially preoccupied with risk; rather, I hold that *every* society is a 'risk society' in the sense to the extent that it emerged and is organised in response to risk. What I have described above is how we experience and cope with technological risk *in our particular way* (a modern way, aimed at more control by means of technology), but this does not mean that we care more about risk than others. This brings me to the next discussion.

4.2.3 Social Technologies and Political Culture

We have always tried to become less vulnerable by building 'communities' and 'societies', that is, by living together in an organised way. Social violence and chaos are transformed into a social order. Sometimes it has been attempted to (re)create a total order, but liberal democracies also depend on, and constitute, *an* order. Inter-human violence is (to some extent) replaced by state violence, and crime is replaced by surveillance and punishment (to use Foucault's terms) to maintain that order. The technological and economic changes described above have changed society, and I will not repeat that analysis here. But there is another dimension I did not mention yet. Society has never been merely passive in relation to technological and economic changes. We have always tried to steer society. We have been political. Moreover, for this purpose we have always tried to manipulate individuals. The history of culture is also the history of manipulation, of social engineering. Foucault has pointed to the various ways modern society has 'disciplined' people (e.g. Foucault 1977). But the premodern religious, technological, and economic orders mentioned before also relied on social engineering—even though not *modern* social engineering. Many means have been employed to reach the goal of social order. 'Politics', then, becomes the steering and management of social order, which uses social (and other) technologies as a means. The motivation for creating social order can be the protection of the individual. This is the modern, contractarian argument. More generally, in order to decrease vulnerability, it seems good for humans to live in communities since that reduces their vulnerability. (Whether or not it is good for humans to consider themselves 'individuals' and to live in *large* communities or societies is a different matter, but this is what has happened.) One way large societies in Europe have reduced individual human vulnerability is by systems of 'social security', human rights, and democracy. However, social orders can also increase vulnerability, for example, by creating a totalitarian state that oppresses people, by allowing market forces to reign largely uncontrolled, or more generally by the absence of policy. Furthermore, by making people dependent on social support, vulnerability has been transformed but not diminished. Many people are now dependent on the state and whatever it does to them. Democracy, moreover, is a highly unstable form of government that can create or allow great uncertainty and chaos. Thus, the problem of social order and vulnerability has not been 'solved', and past attempts to solve it 'once and for all' have increased vulnerability and violence to unacceptable levels.

Thus, society can be understood as (1) a source of vulnerability but also as (2) an anti-vulnerability strategy, which in turn becomes a source of new vulnerabilities, for instance the vulnerabilities related to modern 'mass' societies, individualism, and welfare states.

Is this approach to society too pessimistic? Although it acknowledges the brute conditions of human life, it is not necessarily Hobbesian. It allows room for (unstable) solidarity, trust, and empathy as already belonging to the social. Let me respond to Bryan Turner's neo-Hobbesian argument in order to show this and to explore the relation between vulnerability, sociality, and embodiment.

Turner provides an account that also understands the building of social institutions[7] as a response to vulnerability, institutions which are themselves precarious: they 'cannot provide an enduring, secure and reliable social environment' (Turner 2003, pp. 276–277). But this does not imply that there are only processes of struggle; there are also mechanisms of solidarity and trust—the latter which relies on sympathy and empathy. Turner sees human rights as 'juridical expressions of basic patterns of solidarity' (Turner 2003, p. 277). I will say more about his argument concerning the relation between vulnerability and human rights in the next chapter, but let me first outline Turner's view on the relation between vulnerability, embodiment, and sociality more fully and criticise it in order to clarify and further develop my own position.

Turner sees embodiment as 'the ultimate source of our common sociability' (Turner 2008, p. 242). He explains this as follows:

> Human beings are ontologically frail and their natural environment uncertain. In order to protect themselves from vagaries and afflictions, they must build social institutions (especially political, familial, and ecclesiastical institutions) that come to constitute what we call "society." We need the companionship of society that, through the sharing of bread (pan), provides us with the means of mutual support. (Turner 2001, p. 10; see also Turner 2008, p. 243)

Turner does not make a sharp distinction between sociability and religion: 'We need the creative force of ritual and the emotional effervescence of common festivals to renew social life and to build effective institutions' (Turner 2008, p. 243). However, these social and religious institutions are themselves unstable: 'the world that human beings fashion collectively to form social worlds is inherently and alarmingly precarious' (Turner 2008, p. 244). Turner refers here to Girard. However, he also argues that this precariousness generates 'inter-societal patterns of dependency and connectedness that in their more psychological manifestations result in sympathy and empathy without which society would not be possible' (Turner 2001, p. 10). Thus, solidarity, trust, and empathy arise as a result of our common vulnerability experiences and keep up the social order.

This (dialectic) connection between scarcity and solidarity implies that 'changes to our embodiment must have implications for both vulnerability and interconnectedness' (Turner 2008, p. 242). For human enhancement, this means that technological changes to human 'bodies' are also at the same time vulnerability transformations and social transformations.

I wish to distinguish my view from Turner's on at least two points.

First, Turner suggests that the vulnerability created by social institutions has mainly to do with rituals that 'go awry' (Turner 2001, p. 10), that is, they involve the kind of

[7] Note that *language* too is such a building, a 'house' in which we live and in which we hope to be protected from danger.

violence described by Girard. But I maintain that *modern* institutions as 'juridical expressions of basic patterns of solidarity' (Turner 2001, p. 10), which developed in response to the Girard-type social vulnerability, are also the source of *new* vulnerabilities. For example, as I noted, social security systems can create new vulnerabilities: they create new dependencies. Our legal and political framework cannot give us the immunity we seek, and democracy seems a particularly unstable form of sociality.

Second, Turner's view is explicitly neo-Hobbesian:

> Its premise is that life is nasty, brutish, and short, but, instead of the individualistic notion of a social contract, human and social rights are juridical expressions of basic patterns of solidarity whose foundations are in the common experience of frailty and precariousness, on the one hand, and social interconnectedness, on the other. (Turner 2001, p. 10)

This view is not classical Hobbesian: it emphasises our capacity for sympathy and avoids the notion of a social contract. However, it still presupposes that 'first' there is a state of nature—a state which we might describe as a state of 'natural' risk and 'natural' vulnerabilities—which 'then' is transformed into a society by means of sympathy and solidarity arising from the experience of mutual vulnerability and dependency. Turner's attention to the latter dialectical relation is interesting and constitutes a significant contribution to thinking about vulnerability and sociality. However, according to the existential-phenomenological view I am developing, further steps are needed to go beyond the natural/social dualism. The social has always been there. There is no 'natural' vulnerability apart from the socially transformed vulnerability that marks us at any given time. Risk has always been both social and natural. Human vulnerability is never merely 'natural' but always already 'social'. Thus, we can distinguish at least the following views on the relation between vulnerability and sociality:

1. Vulnerability and sociality are unrelated: the first belongs to 'nature', whereas the second is about the 'social'. The social has nothing to do with our embodiment, and vulnerability is a matter of objective scientific assessment (modern objectivist view).
2. There is a dialectical relation between vulnerability and sociality: the social transforms vulnerabilities, but also creates new vulnerabilities (dialectical view e.g. my use of Turner but also my initial position in this chapter).
3. The vulnerable has always been social, and the social has always been embodied: I moved from the dialectical view (view 2) to what we may call an existential-phenomenological view (view 3), which holds that existential vulnerability experiences are at the same time natural and social.

Note that a somewhat similar non-dualist reply can be made to Beck, who seems to assume that 'first' there is natural (or objective) risk and then modern risk, which is of an altogether different nature. While it is true that modernity created new, previously unknown risks and responded to risks in different ways, the history of risk and vulnerability transformation does not start with a kind of 'state of nature' when risks and vulnerabilities were entirely natural, unspoiled by the social. They have always been social from the start. Therefore, we must question Beck's dualist assumptions. Surely, Beck says that 'nature can no longer be understood outside of society, or society outside of nature' (p. 80). But according to him, this is a stage in history, not something which has always been the case. By contrast, from a non-dualist perspective, 'nature' has never been autonomous or pure in the first place. Risk and vulnerability have always been socially defined and transformed. Consider also the contrast he draws between risk society and previous 'class society':

> The driving force in the class society can be summarized in the phrase: I am hungry! The movement set in motion by the risk society, on the other hand, is expressed in the statement: I am afraid! The commonality of anxiety takes the place of the commonality of need. (Beck 1992, p. 49)

This distinction seems to presuppose that vulnerability can be divided into, on the one hand, 'natural' needs which constitute true sociality and, on the other hand, specific anxieties created by modern risk society and its technologies, of which it is highly doubtful if they can constitute a sociality at all. But this denies that today's anxieties are also rooted in existential embodied vulnerability, which always constitutes some form of sociality. Those who in previous times said 'I am hungry' were also anxious and cared about their safety (otherwise, they would *always* have revolted against those who oppressed them), and those who say today 'I am afraid' are also still needy and vulnerable (otherwise, they would not be afraid to lose their life and/or their comfort), even if their needs and their 'hunger' have been transformed by the modern techno-economic order. The person who was hungry faced 'food insecurity' as well as other kinds of insecurities, and the person who today feels insecure may have enough food but has other needs which he/she may regard as 'basic', and if she cannot count on the food's safety, his/her bodily needs are threatened as well. Hunger is a matter of insecurity and anxiety as much as it is a matter of bodily needs. And the anxiety of the Western consumer-citizen presupposes her embodiment. Vulnerability cannot be split into a purely natural and a purely social or technological component. As an existential condition, its nature and its transformation are at once natural, social, and technological. While different vulnerability transformations may constitute different forms of sociality, all forms of sociality are, as Turner would agree, firmly rooted in our embodiment. The new, late modern risks Beck describes would mean nothing outside concrete, embodied human vulnerability experiences and transformations. Why would I be anxious at all if I believed that radiation, toxins, and viruses could not hurt me? And how could these new risks be described without reference to the health/illness and needs of the human body? Furthermore, living in today's world we may need other things than food, such as new kinds of (and the right kind of) information, and our anxieties may be related to those new needs. New kinds of 'hunger' correspond to new kinds of anxieties. These are not merely 'individual' but shared within a particular vulnerability culture.

If this is true, we have reasons to be optimistic: there are still common vulnerability experiences and anxieties, which means that today there is still a basis for sociality, sympathy, and solidarity, although as I said specific outgrowths of this sociality and solidarity create new dependencies. The modern project of eradicating vulnerabilities, which has led to what is perhaps the most spectacular social transformation of risk and vulnerability we have witnessed in human history (modern technology and the modern state), fails again and again. New risks are being created. If Beck is right that modernisation becomes reflexive, then perhaps now we can achieve some kind of meta-reflexivity and become aware of those failures and of the 'carousel' of technological risk. Then we do not only realise that modernisation creates its own risks which it then tries to manage (and which it created *by* trying to manage); we can also conclude that we have *modernised* our own vulnerability.

4.2.4 Technologies of the Self and Self-Culture

Another important anti-vulnerability strategy that has emerged in the course of the history of human culture is one that concerns the relation to the self. As Buddhist teachings and other ancient wisdom traditions[8] already teach for ages, vulnerability to suffering is (at least partly) due to our own desire(s). When we desire something and we do not get it, we are unhappy. Desire makes us vulnerable to frustration. Moreover, desires can be violent. A dominant current in the history of human culture recommends that we develop self-mastery, take control over our own desires, or, in other (modern) words, that we better establish an inner order. Opinions differ as to which techniques we should use to attain such an order and as to what extent desire should be allowed. But surely we are less vulnerable if we often accept the way things go and control ourselves, rather than desire the impossible, desire more than we can get, and let our desires infinitely increase (the latter seems to be encouraged by our current socio-economic order, boosting our vulnerability). Such techniques that help to reduce vulnerability by mastering desire co-constitute the self. Foucault has proposed the term 'technologies of the self' (Foucault 1976, 1988) when he studied our way of dealing with sexual desire. If I try to master myself, social disciplining is replaced by self-disciplining. The problem, however, is that this strategy has not diminished psychological vulnerability, but only transformed it from a problem of social disciplining and social engineering (Which laws should we have? Should women cover part of their body? How should men behave in public? etc.) to a problem of the self, an inner struggle. Outer repression, which is clear and visible, has been replaced (in the modern West) by inner repression. Freudians might say that this makes us neurotic. And the 1968 hedonist response to this problem has not solved it: we have won outer freedom, but we are left with ongoing and increasing inner desires and frustration. We remain vulnerable or have become even more vulnerable.

In *Either/Or* (1843) Kierkegaard has provided an interesting analysis of the vulnerability of the self and of strategies that try to overcome that vulnerability.[9] Let me interpret his journey of the self through various stages or spheres as a narrative of anti-vulnerability struggle, revealing different options of coping with vulnerability.

[8] The Hindu tradition already understood the human condition as one of suffering due to desire: many desires are never met, which leaves us frustrated, and if they are met, new desires take their place. Schopenhauer was also influenced by this kind of view of the self.

[9] My summary of Kierkegaard on this theme has benefited from Dreyfus and Rubin (1991), who did a good job at explaining the progression through different stages or spheres.

We start in the *aesthetic* sphere, which is the sphere of (immediate) enjoyment. This renders us vulnerable, since enjoyment depends on the body. We may get ill, disabled, too old, and so on, and this will make enjoyment (understood in this way) impossible. We fear that this may happen (and that it might happen rather sooner than later), which makes us even more vulnerable. (This is especially problematic in a hedonistic youth culture, which tries to neglect the possibility of old age and death.)

A possible solution is to reflect on our desires and rely on our imagination to reach invulnerability. We detach ourselves. But this is not real enjoyment since it comes at the expense of losing our connection to the world. We buy invulnerability at the cost of losing enjoyment. We become indifferent. In response to this problem, we may want to move to the *ethical* sphere: we choose to give our life significance. But on what basis can we choose? It seems that we only choose choice itself. If we look for a basis of our choice, we find our desires again.

Kierkegaard's first answer is that we could go for 'religiousness A': we come to see particular needs as relative; our only need is god. Again, this is supposed to render us invulnerable. But this does not overcome the problem of desire Kierkegaard mentioned before: either we are indifferent and therefore invulnerable, but then we have no real desires, and in our mystical contemplation we are disengaged, detached from the world, or we have desires, but then we are vulnerable once again—we are vulnerable to them (the desires) not being satisfied.

'Religiousness B', then, is a stage that is meant to overcome this problem. We commit ourselves to a particular project. This means that we engage with the world and finally give up striving for invulnerability. *We accept risk*. This is what Kierkegaard calls faith. It seems he thought that to seek invulnerability by fleeing to the invulnerable sphere of our imagination or god is a sin; it is a sin to seek a risk-free life. But whatever the normative conclusion drawn, I interpret Kierkegaard's 'religiousness B' as a very particular kind of transformation of vulnerability: desire is still the source of vulnerability, but we are no longer vulnerable to the desire for pleasure not being satisfied. Instead, we are vulnerable to the potential frustration of the particular desires related to our concrete, particular commitment. But we accept that any project is always risky and that we always put ourselves at risk when we commit.

Consider also a further objection to the claim that the realm of imagination or god is a space of invulnerability. Imagination and reflection can even increase desire and its frustration, and as said, to engage into a relationship with a personal god—the belief in the personal Christian god forms the background of Kierkegaard's account—makes us vulnerable to whatever can go wrong in any personal relationship. Just as the more social or communitarian forms of religion subject us to different kinds of risks (having to do with the possibility of communal violence—see previously), the more individualistic forms such as Kierkegaard's Protestantism also merely transform rather than diminish vulnerability. For example, feelings of guilt and sin can torture the self. The merit of what can be called Kierkegaard's account of the vulnerability of the self, then, lies mainly in the insight that self-vulnerability is unavoidable and that for those who follow the existentialist path—which I do not defend here—the task is not to try to escape from vulnerability, but to commit to a particular project and accept the vulnerability that results from this commitment.[10]

I wish to leave it open whether or not this can be done without religion, but if we become religious in the way Kierkegaard meant it, then it cannot be an oasis of invulnerability in the desert storm of desire. It seems that we have to learn to live with the wind-dried skeletons of our unrealised dreams—and with the possibility that it might happen again to our new projects and commitments.

4.3 Conclusion: Vulnerability Transformations as Transformations of a Form of Life

I conclude that with regard to each vulnerability strategy, it first appears that our vulnerability diminishes, but then soon we discover new vulnerabilities, or, rather, we experience that our vulnerability has not diminished but is merely modified, transformed—if not increased. Our vulnerability has been spiritualised and de-spiritualised, globalised, insured, technologised, socially engineered, liberated and repressed, fled from, and accepted, but we remain vulnerable.

This existential vulnerability should not so much be understood as involving a dialectic between two different spheres 'nature' and 'culture', but as being natural and cultural at the same time. Our vulnerability is always already social, and sociality cannot be understood without vulnerability and embodiment.

In order to emphasise that vulnerability transformations are at the same time spiritual, social, technological, and self-shaping, we could borrow Wittgenstein's term 'form of life'. Vulnerability transformations are not merely bodily, not merely social, not merely technological, and not merely psychological. Rather, they are all of that simultaneously: they are changes to our form of life—a specific form of life here and now, different from past and future, and different from what people do in other places.

So far, I discussed *human* vulnerability. Could this change in the future? Is it already changing today? Are we moving towards a posthuman form of life and hence posthuman vulnerability? Some would argue that we are now awaiting transformations of vulnerability that are more radical and more fundamental than ever before—transformations which render it more plausible to speak at least of a *trans*-human vulnerability (by which I mean the vulnerability of humans in transition) if not *post*human vulnerability. In one sense, being human is always being trans-human in the sense discussed: our vulnerabilities are being transformed in various ways. However, discourses about the posthuman seem to suggest that there

[10] Note that by commitment Kierkegaard does not mean a *choice*, then we would remain in the ethical sphere—this seems to be the case for Sartre's idea of radical choice. Rather, commitment seems to have the mixture between activity and passivity Kierkegaard has tried to retrieve from the ancient world. But its precise meaning remains somewhat unclear.

may be a radical break with the past, a fissure in the history of vulnerability. I have mentioned this possibility in Chap. 2, and I will further discuss it in the next chapters.

Leaving aside the desirability of such a posthuman future and the conditions of possibility of *evaluating* such a future (I will say more about this in Part II), we should not neglect significant changes that are already happening here and now, changes to human being and to human vulnerability that are more 'enhancing' than we may expect.

First, as I already mentioned *medical science and technology* has significantly increased our lifespan. Are we today the same beings compared to 500 years ago, given that we now live so much longer? If our existential vulnerability is related not only to mortality as such but also to our experience of the time until death, what has this increased lifespan done to our experience of our lives? We already live different lives now than our grandparents. How did our vulnerability change? How do we experience our vulnerability today as compared to the past, and how will we experience it when medical science and technology further increase our lifespan?

Second, we should not be blind to the anthropological and vulnerability transformations set in motion by *information and communication technology* (ICT). The ICT revolution of the past decades is changing the way we live our lives, the way we relate to others, and the way we see ourselves. If our minds are extended via the Internet, are we still the same kind of beings as our ancestors? If young people experiment with their identity not only at school and on the streets but also in virtual worlds and via online social networks, do they still grow up to be the same kind of humans as their parents or grandparents? What will toy robots do to children's development? What is the use of smartphones, iPads, and other electronic equipment already doing to them now? Our lives are already changed and are still changing; it seems likely that human being and human vulnerability will as well.

Are these exaggerations? Even if the change is smaller than I suggest here, new technological transformations of vulnerability need our special attention today since they raise the normative, existential-ethical question: how should we live our lives? How should we exist? Answering these questions involves answering the question: what should the human become? This problem and the question how to approach it will be the theme of the next chapters.

References

Beck, Ulrich. 1992. *Risk society: Towards a new modernity*. Trans. M. Ritter. London: Sage Publications.

Bostrom, Nick. 2005. The fable of the dragon tyrant. *Journal of Medical Ethics* 31(5): 273–277.

Bourdieu, Pierre. 1990. *The logic of practice*. Stanford: Stanford University Press.

De Mul, Jos. 2006. *De domesticatie van het noodlot: De wedergeboorte van de tragedie uit de geest van de technologie*. Kampen: Klement.

Descartes, René. 1637. *Discours de la méthode*. Trans. L.J. Lafleur. *Discourse on method*. In *Discourse on method and meditations*. Indianapolis: Bobbs-Merrill, 1960.

Dewey, John. 1929/1951. Experience and nature. In *The later works 1925–1953, vol. 1: 1925*, ed. J.A. Boydston. Carbondale/Edwardsville: Southern Illinois University Press.

Dreyfus, Hubert L., and Jane Rubin. 1991. Kierkegaard, division II, and the Later Heidegger. In *Being-in-the-world: A commentary on Heidegger's being and time, division I*. Cambridge, MA/London: The MIT Press.

Foucault, Michel. 1976. *La volonté du savoir.* Trans. R. Hurley. *The history of sexuality*, vol. I. London: Pelican Books, 1981.

Foucault, Michel. 1977. *Discipline and punish: The birth of the prison*. Harmondsworth: Penguin.

Foucault, Michel. 1988. Technologies of the self. In *Technologies of the self: A seminar with Michel Foucault*, ed. L.H. Martin, H. Gutman, and P.H. Hutton. Amherst: The University of Massachusetts Press.

Girard, René. 1972. *La Violence et le sacré*. Paris: Bernard Grasset. Trans. P. Gregory. *Violence and the sacred*. Baltimore: The Johns Hopkins University Press, 1979.

Kierkegaard, Søren. 1843. *Either/or: A fragment of life*, vol. 1. Trans. D.F. Swenson and L.M. Swenson. Princeton: Princeton University Press, 1944.

Merleau-Ponty, Maurice. 1945. *Phénoménologie de la perception*. Paris: Gallimard. Trans. C. Smith. *Phenomenology of perception*. Abingdon/New York: Routledge, 2002.

Turner, Bryan S. 2001. The end(s) of humanity: Vulnerability and the metaphors of membership. *The Hedgehog Review* 3(2): 7–32.

Turner, Bryan S. 2003. Biology, vulnerability and politics. In *Debating biology: Sociological reflections on health, medicine and society*, ed. S.J. Williams, L. Brike, and G.A. Bendelow. London/New York: Routledge.

Turner, Bryan S. 2007. Culture, technologies and bodies: The technological Utopia of living forever. *The Sociological Review* 55: 19–36.

Turner, Bryan S. 2008. *The body and society: Explorations in social theory*. London: Sage.

Part II
Normative Anthropology of Vulnerability

Chapter 5
Ethics of Vulnerability (i): Implications for Ethics of Technology

5.1 Vulnerability and Ethics

In Part I, I have outlined an existential-phenomenological approach to risk and vulnerability according to which we are always already being-at-risk, rendering ourselves vulnerable by engaging with the world. I have also argued that in our struggle against vulnerability, we create new vulnerabilities and thereby transform ourselves as much as we transform the world. Now I want to show that this relational anthropology of risk and vulnerability has important implications for ethics of risk and the evaluation of new technologies. We need a *normative anthropology of vulnerability* which does not ask which objective risks are acceptable but what the human should become—that is, which vulnerability transformations we want.

What does this mean? In this chapter, I will take the first steps towards a normative anthropology of vulnerability by exploring the idea of an 'ethics of vulnerability' and what it would imply for ethics of technology. Therefore, this chapter will be mainly asking questions, in particular questions concerning the approach I propose. In the next chapters, I will offer more 'content' in the sense that I will not only continue my outline of the approach but also provide reflections on the posthuman future and on vulnerability in the information age.

5.1.1 The Value of Vulnerability and the Vulnerability of Value

What is the relation between vulnerability and ethics? Surprisingly, there are 'only passing references to human vulnerability' in the history of ethics (MacIntyre 1999, p. 1). This neglect of vulnerability even includes much of contemporary medical ethics and bioethics, where one would expect vulnerability to be an important concept. As Dell'Oro notes, 'in the index of what is arguably the most influential work and the most widely used approach to bioethics in America, *Principles of Biomedical Ethics*, by Beauchamp and Childress, vulnerability is not even mentioned'

M. Coeckelbergh, *Human Being @ Risk: Enhancement, Technology, and the Evaluation of Vulnerability Transformations*, Philosophy of Engineering and Technology 12,
DOI 10.1007/978-94-007-6025-7_5, © Springer Science+Business Media Dordrecht 2013

(Dell'Oro 2006, p. 3). But although vulnerability is usually not explicitly discussed, it is not absent from philosophical discourse altogether. In some twentieth-century ethical theories, ethics is understood as a response to vulnerability, and in this sense, vulnerability has played the role of a kind of *foundation* or *necessary condition* for ethics. Let me explain this and give some examples.

First, the concept of human rights can be interpreted as a response to human vulnerability. Human rights are responses to human vulnerabilities in the sense that for every human right, there is a corresponding vulnerability and risk, the risk of the potential violation of the right. Many of these rights are a response to the vulnerability that arises from the possibility of state violence; later other kinds of rights have been added, corresponding to, for example, social and economic vulnerabilities. As a normative concept, human rights are employed to reduce vulnerability or, at least, to justify its reduction. Bryan Turner has argued that human rights can be grounded in a concept of human frailty (especially the vulnerability of the body), in the idea of the precariousness of social institutions, and in moral sympathy. Human vulnerability can be contained or ameliorated by rights (Turner 1993). In a similar way, Nussbaum and Sen's 'capability' approach (Nussbaum and Sen 1993; Nussbaum 2006) is a response to vulnerability. Supporting people's capabilities can be interpreted as efforts to reduce the corresponding vulnerability and risk. If my capability of health is promoted, then I am less vulnerable to disease. If my capability of social affiliation is enhanced, then I am less vulnerable to loneliness or social violence.

Note that Turner writes as a sociologist who takes seriously biological accounts of human nature. He criticises relativistic social 'constructionism' and the so-called cultural turn for seeing biology as 'merely a rhetorical device' (Turner 2003, pp. 272–273). For example, he notes that Marx had a view of the natural condition of human beings, which he needed for his theory of alienation. Therefore, Turner criticises Foucault for seeing the body as 'a contingent effect of power rather than a fact of nature' (Turner 2003, p. 275).

With his focus on biology and the body, Turner is in tune with contemporary sociologists and philosophers who emphasise 'embodiment' and 'the body' in their theories—without, however, turning 'biology' and 'the body' into a social-cultural construct (alone). Reliance on the concept of 'the body' is fashionable, and I sympathise with Turner's rejection of (this version of) social constructivism. However, we must be careful not to reduce vulnerability to its biological dimension. For Turner, vulnerability is mainly a property of the body, a 'fact of nature'—even though he allows for embodiment as a *social* process that 'disciplines our biological foundations'[1] (Turner 2003, p. 277) and recognises that embodiment has a subjective dimension.[2] In the view I am articulating in this book, vulnerability arises out

[1] Note the influence of Foucault here (Foucault 1977).

[2] Turner distinguishes three dimensions of embodiment: 'one can talk about having a body in which the body has the characteristics of a thing, being a body in which we are subjectively engaged with our body as a project, and doing a body in the sense of producing a body through time' (Turner 2003, p. 281). With Merleau-Ponty, we can say that the body is not only an object (*Körper*) but also lived experience (*Leib*). However, I insist that the concept of 'embodiment', even if broadly conceived in this way, is still not rich enough to do justice to the relational nature of vulnerability.

of the relation between us (including our biological, bodily dimension) and the world. In this way, I hope to avoid both (reductive) naturalism and (radical versions of) social constructivism. The biological, the social, and the technological do not exist in themselves but are aspects of our way of experiencing (passive) and coping (active) with the world. Our ways of being-vulnerable and our ways of coping with vulnerability are part of our mode of existence.

This concept of existential vulnerability also allows me to achieve one of Turner's (other) aims. He objects to Foucault's attention to 'how social practices are inscribed on the human body as a merely passive object' and proposes instead more emphasis on 'the lived body' (Turner 2003, p. 275). I believe Foucault's later work on *technologies of the self* (Foucault 1988) does more justice to the active dimension; Turner neglects this later work. Nevertheless, if Turner's point is that we should account for the active and passive side of living bodily experience, I agree. My approach achieves this aim by emphasising the active aspect of the emergence of vulnerability and by calling attention to the *experience* of vulnerability rather than to biological (or sociological) 'facts'.[3] Vulnerability involves lived activity: it only takes shape by means of our imagination, interpretation, and emotions and by means of an active engagement with the world.

Second, Levinas's ethics starts from human beings as vulnerable beings. But it shifts the focus from *my* vulnerability to the vulnerability of the *other*. The naked, vulnerable face of the other appeals to our responsibility: it asks 'do not kill me'. In this sense ethics is prior to cognition. First there is the encounter with the face of the other, that is, first there is the human relation and in that relation I have to respond to the vulnerability of the other. I am called by the other to respond (Levinas 1961). Levinas sees this as something passive, something that happens. It is an interruption. In *Humanisme de l'autre homme* (1972), he writes: 'That the other is my concern, happens against my will' (Levinas 1972, p. 82). The other is always *already* standing under my responsibility. In this view, ethics has to do with an inability ('cannot') rather than an 'ought'.[4] The vulnerability of the other holds my consciousness in a grip. In this sense, responding to (the other's) vulnerability is what ethics is about.

Third, in ethics of medicine, health care, and education, care is understood as a response to human vulnerability, for example, the vulnerability of children or patients. Here cultural, anthropological, and phenomenological perspectives are interesting, since they go beyond 'ethics' narrowly conceived as 'applied ethics',

[3] This is also why vulnerability cannot be measured, in contrast to what Turner suggests. Of course we have 'indices of disease, disability, chronic illness, morbidity, life expectancy' (Turner 2003, p. 280), etc., but such statistics do not reflect people's *experience*. An empirical study that seeks to complement a philosophical inquiry into vulnerability must not only take into account statistical and other 'data' but should also look at other approaches such as narrative accounts of lived and personal experience. (See for example Jackson's phenomenological anthropology as referred to in Chap. 3.)

[4] Hannah Arendt, too, makes this point in her later work: ethical responsibility is related to my inability to do something or to live a certain life: I *cannot* do it. More precisely, it is related to my inability to *imagine* acting such and living such a life.

beyond the application of ethical theory (principles, norms, etc.). For example, Mark Nichter defends an anthropological approach to medical risk: 'What are needed are finer grain assessments of how individuals actually cope with states of vulnerability and perceptions of risk' (Nichter 2003). And Kay Toombs discusses three manifestations of vulnerability in illness. Let me summarise her analysis in order to illustrate her approach.

Toombs offers a phenomenological account of illness and the vulnerabilities that come with it. When one discovers that one is ill, there is the perception of bodily threat: one is at the mercy of one's own body, feels alienated from one's own body; the body is no longer taken for granted. There is a loss of control. And when one discovers that one's illness cannot be cured, this is the ultimate loss of control. Here the imagination of the future plays an important role: the threat is not immediate, but 'the perception of *future* threat often shapes the patient's *present* experience of illness' (Toombs 2006, p. 114). Second, illness disturbs one's relations with the world. For someone with a bodily disorder, the world resists; it is no longer familiar but a threat, or a challenge. Third, personal integrity is threatened. Patients find it difficult 'to retain a sense of self-worth in the face of cultural attitudes that place inordinate value on unrealistic ideals of autonomy, independence, productivity, health, and beauty and that view illness, suffering, and death as meaningless affronts to human dignity' (Toombs 2006, pp. 126–127). If such cultural attitudes dominate, social death precedes physical death (Toombs 2006, p. 130). Toombs claims that someone who is ill and (therefore) already experiences vulnerability suffers even more if she lives 'in a world where success is measured by the ability to shield ourselves from vulnerability, and status is marked by the degree of one's insulation from potential harm' (Toombs 2006, p. 119). Using the vocabulary I offered before, we might say that the illness makes the person not only physically vulnerable but also socially. Against the dominant cultural attitudes, Toombs argues that ethics must acknowledge shared dependency and *embrace* vulnerability (and goes on to defend a Christian ethics that does as much).

Fourth, as this last remark suggests, vulnerability can also be good, or can at least be seen in a different light. Ethics of care, by which I mean the specific theoretical approach rather than a subfield of ethics, is especially interested in vulnerability given its critique of Kantian and utilitarian ethics. Both Kantian and utilitarian ethics are, in the words of Dell'Oro, 'joined by similar representations of the moral subject as autonomous, individualistic, and self-sufficient' (Dell'Oro 2006, p. 3). In ethics of care, by contrast, vulnerability is emphasised as opposed to autonomy and self-sufficiency. Some ethics of care theories are influenced by phenomenological and neo-Aristotelian 'human flourishing' perspectives on human being. They all share the view that vulnerability is, or can be, something positive, a good. In neo-Aristotelian approaches, it is even seen as *necessary* for human flourishing. MacIntyre asks us to 'treat the facts of vulnerability and affliction and the related facts of dependence as central to the human condition' (MacIntyre 1999, p. 4), although he overemphasises bodily vulnerability and refrains from a broader analysis such as the one proposed in this book. According to Frohmann, influenced by MacIntyre, we need virtues that are 'informed by a rational knowledge of bodily disability and vulnerability, and the animal nature of our human condition' (Frohmann 2000, p. 427). Marian Verkerk—who sees care ethics as a moral stance—has put forward a critique of the ideal of independency in human life. Instead of autonomy, she stresses other values such as trust, caring, and responsibility. She does not reject autonomy but proposes a relational account of it: as a moral capacity, autonomy 'can only be developed in relation to others' (Verkerk 2001, pp. 291–292). She is influenced by Margaret Walker, who emphasises dependency, vulnerability, and social interdependence. And Alisa Carse has criticised what she calls 'the myth of the in-control agent', which ignores much about the human condition—in particular vulnerability (Carse 2006, pp. 35–36). Our flourishing depends on others. She concludes that 'allowing ourselves to be *vulnerable* is necessary to loving and being loved, to caring and being cared for, and to playing, exploring, and growing in ways that strengthen and vitalize our effective agency. Our vulnerability is inextricably tied to our capacity to give of ourselves to others, to treasure and aspire, to commit to endeavours, to care about justice and about our own and others' dignity' (Carse 2006).

Twenty years earlier, Nussbaum already argued in *The Fragility of Goodness* (1986) that the good life partly depends on vulnerability. In her preface to the revised edition, she says that the invulnerable life 'is likely to prove impoverished' and that vulnerability is 'a necessary background condition of certain genuine human goods'—although she stresses that she did not endorse the view that vulnerability is to be prized in its own right or as an end in itself (Nussbaum 1986, p. xxx). She gives the example of love of children (a genuine good): if we love a child, we make ourselves vulnerable. She also says that tragedy can 'enrich our sense of how human values are vulnerable to chance' (Nussbaum 1986, p. xxix).

While I sympathise with most of these approaches to ethics, I wonder how helpful their views on the relation between ethics and vulnerability are for the project of a normative anthropology of vulnerability. I already offered some objections above, but let me now provide more systematic criticism.

The foundational role vulnerability plays in human rights theory, in Levinas, and in ethics of care is highly problematic, since vulnerability is not in itself normative (Kottow 2004). I believe this is right in at least three ways.

First, while it is plausible that 'the growth of human rights legislation, institutions and culture in the twentieth century was based on a common recognition of human vulnerability' (Turner 2003, p. 278), this does not 'ground' them in vulnerability, in the sense that vulnerability alone does not *justify* human rights. The history of the recognition of our common vulnerability only *explains* how the idea of human rights could gain so much support.

Second, even this claim is not entirely convincing, since human vulnerability itself has proved insufficient to *motivate* ethical behaviour and to prevent violence. The passivity Levinas speaks of only makes sense if 'something else'—for example, a justification or ethical attitude or habit independent of vulnerability itself—motivates us to act ethically and to 'see' the face of the other in the first place. The face of the other only speaks if I already have taken the starting position required for an ethical attitude and ethical behaviour (in the case of Levinas, this was the position put forward in the Jewish-Christian ethical tradition) and if I am already socialised, if I am raised in a society, if I have gone through a process of social-moral development, and if I have the right kind of social-moral habit. My relation to others and their vulnerability is already mediated and shaped when a particular encounter happens. Similarly, our capacity for sympathy appealed to by Turner and Nussbaum is, by itself, insufficient to prevent violence (Turner 2003, p. 279). Unfortunately, the 'social glue' is not strong enough.

Third, human rights theories focus so much on the victim, that the source of violence remains largely out of sight. Such theories are misguided by their conception of vulnerability as a feature of the world *or* as a feature of me or the other (as the not-me), rather than something that only emerges in a relationship between me and 'that what may harm me'. But even acknowledging relational vulnerability and assuming that people have the required motivation is an insufficient grounding for ethics for two reasons: (1) it seems that we need an independent, additional justification for acting to diminish vulnerability, and (2) even if such a justification is not necessary, it may be unclear what we should do to reach that aim. Even if I am called to respond to the other's vulnerability and *want* to respond (feel obliged to respond, feel responsible), it may be unclear *how* best to respond; this is at least *also* an ethical question. Thus, vulnerability alone is not *sufficient* for grounding ethics—whatever else may be needed to ground it.[5]

The neo-Aristotelian view, then, rightly understands vulnerability as a *necessary* condition for 'ethics' (but not a foundation or sufficient condition). If ethics concerns 'the good life' and 'human flourishing', then vulnerability can be good, or contribute to the good. According to MacIntyre, acknowledging vulnerability—in particular dependence on others—is a virtue and is necessary to live the good life. In my conceptual framework, I put it in a different way. As I noted in Chap. 3, existential vulnerability can also be defined as a precondition of the possibility of the good: our existence carries within it the possibility that the good emerges. Vulnerability is a precondition for the good to emerge—albeit perhaps not a sufficient one. And human rights and ethics of care approaches can rightly be understood as responses to human vulnerability.

However, in their mission against an impoverished view of personal autonomy, ethics of care theorists tend to overemphasise the passive side of vulnerability, what *happens* to the receiver of care (the experience of loss of control, for example.), and hence risk downplaying the strategies that that person and those who care for her can use to reduce her vulnerability. Furthermore, although I sympathise with MacIntyre's claim that we should acknowledge vulnerability and dependence, his view relies on an essentialist definition of our 'human nature' as dependent, rational animals. In addition, his use of the term 'the good' suggests a conception of good as something external to human experience and existence (e.g. a neo-Platonic Oneness or something of divine origin or linked to the divine). My approach to vulnerability and to the discussion about human

[5] My objections need not imply moral relativism in the sense of a belief in the absence of any ground for ethics. I leave the question concerning the foundation of ethics open at this point.

enhancement and new technologies does not depend on neo-Aristotelian essentialism, Platonism, or Christian (in MacIntyre's case, Catholic) ethics.

5.1.2 *Evaluating Vulnerability Transformations*

My own view on the relation between vulnerability and ethics, and on what should become an ethics of vulnerability as a normative anthropology of vulnerability, does not start with a discussion of ethical theory, but with an analysis of vulnerability. Normative anthropology needs descriptive anthropology, without being *based* on it in the sense of justification. Let me explain how I wish to proceed. In the previous chapter, I have discussed how vulnerability is transformed by technology and by other anti-vulnerability strategies. Now ethics can be seen as such a strategy (a response to vulnerability) or as a theory about the good that partly relies on vulnerability rather than being only opposed to it, but I will understand ethics—the ethics of vulnerability—as the normative evaluation of anti-vulnerability strategies, taking into account ongoing vulnerability transformations. In this sense, vulnerability is part of the object of ethics, but not its foundation or its principle. And such an ethics is a response to human vulnerability, but not a direct response (which it is for Levinas, ethics of care, or human rights theories). It concerns *reflection* about *how* we should respond.

Let me explain this by clarifying how I believe ethics of technology should be approached. Understood as applied ethics, ethics of technology is about applying well-known ethical principles to problems and cases that involve technology. Understood as an ethics of vulnerability, however, ethics of technology starts with a descriptive stage: how does a specific technology alter our vulnerability? Then one must evaluate whether that transformation contributes to the good life, to human flourishing. Thus, it connects the micro-perspective (I am confronted with vulnerability of the other and respond to it ethically) with a macro-perspective: this is the technological strategy *we* use to cope with vulnerability; these are our vulnerability habits, so let us ethically evaluate these strategies in terms of their contribution to the good life. The descriptive and the normative go hand in hand. The starting point of this ethics as normative anthropology is how technology changes human being (descriptive anthropology). In the same way, we can evaluate religious, financial, economical, and psychological vulnerability changes.

Let me show what we can gain by this kind of ethics by briefly discussing vulnerability questions raised by some new or emerging technologies in the domains of robotics (in particular personal robots) and medicine/biotechnology (in particular genetic enhancement).

5.1.2.1 Personal Robots

There is a trend in robotics from industrial applications towards personal robots: robots that play a role in personal life and 'share physical and emotional spaces with the user' (Cerqui and Arras 2001). Consider, for example, household jobs, entertainment, sex, health care, and companionship. In order to ethically evaluate this near-future scenario, there are at least two possible approaches. One is to do 'applied ethics'. For example, we may wish to apply Kantian ethical theory to the question of health care robots and ask, for instance, if replacing nurses (humans) by care robots respects a patient as an end in itself, rather than a means to, say, a 'financially healthy' health care system. Another approach is to start from the ethical-anthropological question: How would such robots transform our lives and our way of being? The normative, evaluative moment is postponed. It is impossible to decide the question of the good life (or, for that matter, another question asked by another normative-ethical theory) *a priori*, that is, without first imagining how it would be to lead such a life. In the case of the introduction of robots in the personal sphere, this approach implies that we should use empirical and/or phenomenological research to know more about current and near-future human-robot interaction and its context. How do humans interact with, for example, pet robots, and how does this change their experience, their engagement with the world, their existence? In particular, how does this interaction transform human vulnerability? For example, what is the good life for people who are living in a care institution for elderly people? How do they experience their vulnerability, and how may this be changed by care robots? This also means that we should ask more specific questions as well as general ones. How does person X experience her vulnerability in situation A? How does robot R influence person Y's relation to the world? What are the fears and hopes of person Z, and what can the technology do in relation to them? Including such questions is likely to yield a richer analysis of possible ethical problems than restricting the analysis to the application of a priori principles. Such principles can still play a role in the analysis as explicitations of what we value, but they cannot be the starting point or the only normative reference point. Moreover, we must take seriously the ethical importance of appearance. For the person's existential experience, it matters little whether or not the robot involved is 'really' intelligent, or 'really' has a mind that more closely resembles that of humans. What matters is what that robot does to the human person in the context and situation of particular interactions. For instance, if the robot appears like a dog and acts like a dog, this appearance matters for the social-emotional experience of the human person. Note also that what happens in the interaction will also depend crucially on the character and experience of the person involved. For example, if I do not like dogs or even fear them, it is unlikely that a dog robot will do much to enhance the quality of my existence.

However, when it comes to evaluating future scenarios, we face the following problem. We use our imagination. Imagination is not necessarily fantasy. Moral imagination is not free-floating, but rooted in knowledge. But what kind of knowledge can we have of the future? To the extent that personal robots are not yet used, it may be difficult to imagine the exact vulnerability changes. Furthermore, at this point in time we may not know the precise users of the technology and their context. My short answer for now is that we can at least benefit from existing experiments and research while acknowledging that our ethical-imaginative efforts have their limitations. My longer answer follows on the next pages and in the next chapter.

5.1.2.2 Human Genetic Enhancement

Let me further clarify my approach by discussing another possible near-future technology: genetic enhancement. As suggested above, in order to evaluate vulnerability changes, we must first know more about human vulnerabilities and their transformations in a specific case. These are not fixed. Moreover, it is not obvious to take up this task since most ethical assessments do not consider vulnerability as a theme, do not understand it as a *relational* property, and discuss 'facts' rather than appearance and experience.

Let me analyse Nick Bostrom's discussion of human genetic enhancement (Bostrom 2003) to show the gains of an ethics of vulnerability applied to issues concerning genetic enhancement technology and to point to a major problem with imagining future vulnerability changes. First, the project of transhumanism directly aims to diminish or eliminate human vulnerability by extending the human healthspan (a long and healthy life as opposed to merely a long life), eradicating disease, and eliminating unnecessary suffering (Bostrom 2003, p. 493). For transhumanists, genetic engineering should contribute to these aims. The potential gains mentioned by Bostrom include being free from severe genetic diseases and having a more robust immune system. Thus, the target is—put in my terms—to diminish the vulnerability that arises in the relation between me (living, doing things) and genetic or other sources of disease (phenomenologically, both are experienced as alien, not-self, strange intruders in my life). If human enhancement were limited to this kind of improvements, it seems that its difference from previous medical transformations in vulnerability would not be categorical, since we already managed to reduce risk of disease in the course of the history of medicine. If other conditions are met, such as equality and justice, there seems to be no ethical problem. There *is* a problem, however, if the transformations were to be so profound that we would no longer know 'disease' as we know it today. Such transformation would change the meaning of what is to be human. At least, it is problematic from our contemporary point of view. We may judge this change as ethically problematic on the basis of our current values. For instance, we value health, but we also accept that we can be ill sometimes. If this contingency is replaced by a contingency of a totally different kind, perhaps one having to do with possible mistakes in a laboratory or errors in computer programmes, then 'disease' is no longer 'disease', and human being changes. Furthermore, what if our bodily experience changes radically? In those cases, it also seems that we have no choice but to evaluate whether or not this change is desirable or acceptable by using our *current* ethical standards and values, the knowledge that is available to us *today* for judging what constitutes the good and flourishing life.

The following problems emerge. First, how can we imagine the precise vulnerability transformations that could or will take place? This was already a problem with regard to near-future scenarios involving personal robots, but one could argue that a reliance on empirical research (i.e. current experiments with new robots) could fill the gaps as far as the near future is concerned. The remote future, however, seems to pose a more serious problem. Our society and our ways of living may change dramatically. Second, if human being changes, the values by which we evaluate transformations of vulnerability may change as well. Is the ethics of vulnerability I propose itself *dependent* on its object, on the very processes it tries to evaluate from an 'external' point of view? How 'external' is it given my emphasis on personal experience? How much distance should we take in ethics, and how much distance *can* we take? What if human being (and society) changes in such a radical way that our values, our ideas of the good life and of human flourishing, change as well? This problem is pertinent if we consider possible technological transformations of vulnerability in the remote future (assuming we want our ethics to be able to respond to such a challenge), and demands us to further inquire into the *possibility* of a posthuman ethics, which I shall do in the next chapter.

5.2 Ethics of Technology as an Ethics of Vulnerability

Before I further discuss the possibility of a posthuman ethics, however, let me clarify the conception of technology involved in the ethics of vulnerability, in order to distinguish it from standard approaches to ethics and to technology.

5.2.1 Standard View: Human Values Versus Technological Means

Standard ethical theories assume a conception of technology that is value neutral. Technologies are the means by which we realise human ends. According to this view, what matters in ethics is the evaluation of the ends: we have to make sure that the ends are good and that technologies are used for reaching these ends rather than others. Evaluating new technologies, then, involves finding out if a technology will be used for good purposes, recommending that it must be used for good purposes, and perhaps prohibiting use for bad purposes. For example, in this view nuclear technology is itself neutral—all depends on who uses it for what purpose. Thus, a sharp distinction is made between the sphere of ends and the sphere of means, which goes together with a sharp distinction between humans and non-humans. Values belong to the human sphere; things are instruments for reaching those values.

For discussions about technological risk, this approach implies that an *ethics* of technological risk—indeed an 'ethics of technology'—is *in a strict sense* impossible. If ethics is only about what humans do and what humans value, then the choice of the means is not really an *ethical* decision but a matter of instrumental reason: choose the technology that is the least risky and that is most effective. According to this view, technological risk cannot be *ethically* evaluated, since this would imply that the technology is not neutral. Risk can only be 'assessed'. It is a technical decision we better leave to experts. Ethics and politics should set the aims; it is not their business to say something about what means are fit to reach those aims.

Moreover, in this picture *vulnerability* is neatly divided up between the human sphere and the technological sphere. Humans are vulnerable, and technological systems are vulnerable, but there is no intrinsic connection between the two. If something goes wrong in a technical installation, humans may be 'impacted' by the event, but the technology and the technological risk are seen as external to the humans.

According to this view, responsibility is an entirely *human* affair. The *user* of the technology is responsible when something goes wrong. If the user counts as a moral agent, then the user is responsible. What did the user decide? What were her intentions? What are the consequences for others? With a set of principles, we can judge what the human user has done (wrong). When an oil disaster happens and we want to ascribe responsibility, for instance, an explosion on an oil production installation, we need to track down the decisions of human users of the technologies that resulted in the disaster. If an airplane crashes, we need to find out if the human pilots did something wrong.

This view also assumes that ethics remains the same over time: as a control mechanism for the human mind, with no intrinsic connection to the world, it does not change as long as the human mind remains roughly the same. Therefore, this view is meta-ethically conservative in relation to technology: when new technologies are introduced and used, we do not need a new ethics. We need, as always, better people.

5.2.2 Alternative: Learning to Be-at-Risk

By contrast, the existential-phenomenological approach to vulnerability and the related ethics of vulnerability I wish to develop here challenges these assumptions. It does not view technology as ethically neutral and as disconnected from human being but rather as part of what ethics should deal with and as part of what we are as humans. The philosophical-anthropological concepts 'being-at-risk' and 'being-vulnerable' proposed in Chap. 3 assume a relational view of the human, according to which human being cannot be defined without technology and vice versa. Our vulnerability is co-constituted by the technologies we use, and technology is at the same time part of the way we cope with vulnerability. There is no human vulnerability opposed to, or separate from, technological vulnerability. Humans are 'technologically' vulnerable, and technologies are 'humanly' vulnerable. Technological risk is not something external but part of human activity (by which we put ourselves at risk), and human vulnerability is not external to the vulnerability of technological systems. Both are interdependent and intertwined.

According to this view, ethics is not about human values and principles isolated from human-technological practice but is as much an ethics of humans as it is an ethics of technology. An ethics of vulnerability is about humans-in-relation-to-the-world, not about human minds (e.g. about intentions) and 'external' consequences.

Vulnerability is human-technological. Responsibility is human-technological. What matters is not so much how we best realise human values (value as something 'out there') but how we can better relate to the world (value in action). It is not only about the responsibility of users, since strictly speaking there are no longer users. 'Users' are co-designers. What technology is and what we are changes, and humans and technology change one another. In the process, we are hopefully learning how to be-at-risk in a good way, in a better way. New technologies mean new worlds, new humans, and new ethics. There is no external criterion. To ask the ethical question is to ask how we can cope with the world and with risk in a better way. We cannot become invulnerable. There is no 'outside' to risky and vulnerable existence. But as I will argue in the next chapter, in the future we might end up with different vulnerabilities and different vulnerability configurations and hence with a different kind of human being and human condition—if still 'human' at all.

5.3 The Design and Growth of Human Vulnerability

In the first chapter, I already noted that the human is not entirely open to design and that we have no full control over changes to the human. This is also an important point with regard to human vulnerability. The ethics of vulnerability as an ethics of vulnerability transformations is not an ethics of 'design', if design means that we can develop an entirely new 'concept' human vulnerability, a 'code' or 'blueprint' of a new vulnerability configurations for the human. The space for human agency with regard to vulnerability transformations is limited by natural, social, technological, and other constraints. Therefore, an ethics of vulnerability can guide our coping with technology, risk, and vulnerability only within these limits.

To accept this requires a turn to a less modern understanding of ethics, which understands the human not as separate from the world but as being-in-the-world. Modern ethics assumes that we are external lawgivers or designers (humans as moral agents) who then shape and control the world ('nature', animals, humans as moral patients). The model for this kind of relation is the monotheist God as the creator and manipulator of the world. To use a contemporary metaphor, it is assumed that humans are the programmers and system administrators of the world; they write the world's code, design its sites, and manage its data. From this perspective, an ethics of vulnerability is meant to guide the *making* of human vulnerability, the implementation of a vulnerability 'concept'. But this picture of vulnerability does not recognise that vulnerability transformations are not entirely in our hands. The transformations are always partially 'out of hand', out of human control. On the one hand, it makes sense to say that we 'designed' and 'built' societies, that we transformed the natural world, etc.: humans have a significant impact on their natural and social environment, and arguably in modernity, this impact has increased. But on the other hand, it is also true that in a sense these social, technological, financial, economic, and indeed anthropological transformations have developed 'themselves', that they have evolved and grown. Similarly, the vulnerability transformations related to these changes are not entirely under our control. We can think about which vulnerability changes we want, but we cannot simply *choose* or design our vulnerability configuration. Perhaps the metaphor of *growth* is more adequate than that of design. Human vulnerability has grown into what it is now, and it will grow again into new forms in the future. This growth is only partly the result of human interventions and human 'design'. At most, we are its gardeners. The ethics of vulnerability (transformations) is the ethics of the modest and cautious gardener, who respects the limits of her own powers and understands herself as part of the vulnerability ecology.

Yet in so far as we *have* the possibility to co-shape the form of human vulnerability—and thus the form of the human—what form should this be? And before we can answer that question: what form *could* it be? To what extent can we imagine the future of human vulnerability, indeed the future of the human? What if it changes so radically that we cannot imagine it yet? And what if our *ethics* also changes in the future? Can we step into the same ethical river twice?

References

Bostrom, Nick. 2003. Human genetic enhancements: A transhumanist perspective. *The Journal of Value Inquiry* 37: 493–506.

Carse, Alisa L. 2006. Vulnerability, agency, and human flourishing. In *Health and human flourishing: Religion, medicine, and moral anthropology*, ed. Carol Taylor and Roberto Dell'Oro. Washington, DC: Georgetown University Press.

Cerqui, Daniela, and Kai O. Arras. 2001. Human beings and robots: towards a symbiosis? A 2000 people survey. In: *Post-Conference Proceedings PISTA 03*, ed. Carrasquero, Jose Vicente et al., 408–413 Orlando, Florida. http://www.informatik.uni-freiburg.de/~arras/papers/cerquiPISTA03.pdf.

Dell'Oro, Roberto. 2006. Introduction. In *Health and human flourishing: Religion, medicine, and moral anthropology*, ed. Carol Taylor and Roberto Dell'Oro. Washington, DC: Georgetown University Press.

Foucault, Michel. 1977. *Discipline and punish: The birth of the prison*. Harmondsworth: Penguin.

Foucault, Michel. 1988. Technologies of the self. In *Technologies of the self: A seminar with Michel Foucault*, ed. L.H. Martin, H. Gutman, and P.H. Hutton. Amherst: The University of Massachusetts Press.

Frohmann, Bernd. 2000. Cyber ethics: Bodies or bytes? *International Information & Library Review* 32: 423–435.

Kottow, Michael H. 2004. Vulnerability: What kind of principle is it? *Medicine, Health Care, and Philosophy* 7: 281–287.

Levinas, Emmanuel. 1961. *Totality and infinity*. Trans. A. Lingis. Dordrecht: Kluwer Academic Publishers, 1991.

Levinas, Emmanuel. 1972. *Humanisme de l'autre homme*. Montpellier: Fata Morgana.

MacIntyre, Alasdair. 1999. *Dependent rational animals*. Chicago: Open Court.

Nichter, Mark. 2003. Harm reduction: A core concern for medical anthropology. In *Risk, culture and health inequality: Shifting perceptions of danger and blame*, ed. Barbara Herr and Laury Oaks. Westport: Praeger.

Nussbaum, Martha. 1986. *The fragility of goodness: Luck and ethics in Greek tragedy and philosophy*. Cambridge: Cambridge University Press, 2001.

Nussbaum, Martha. 2006. *Frontiers of justice: Disability, nationality, species membership*. London/Cambridge, MA: Harvard University Press.

Nussbaum, Martha, and Amartya Sen. 1993. *The quality of life*. Oxford: Clarendon.

Toombs, S.Kay. 2006. Vulnerability and the meaning of illness: Reflections on lived experience. In *Health and human flourishing: Religion, medicine, and moral anthropology*, ed. Carol Taylor and Roberto Dell'Oro. Washington DC: Georgetown University Press.

Turner, Bryan S. 1993. Outline of a theory of human rights. *Sociology* 27(3): 489–512.

Turner, Bryan S. 2003. Biology, vulnerability and politics. In *Debating biology: Sociological reflections on health, medicine and society*, ed. S.J. Williams, L. Brike, and G.A. Bendelow. London/New York: Routledge.

Verkerk, Marian A. 2001. The care perspective and autonomy. *Medicine, Health Care and Philosophy* 4: 289–294.

Chapter 6
Ethics of Vulnerability (ii): Imagining the Posthuman Future

6.1 The Possibility of a 'Posthuman' Ethics

A normative anthropology understood as an ethics of vulnerability interprets and evaluates vulnerability transformations. But at the end of the previous chapter, I identified the following problem: how can ethics cope with radical vulnerability changes, and therefore radical changes in our existential condition, that may happen in the *future*? In this section, then, I will discuss the possibility of a posthuman ethics: an ethics that guides, and is adapted to, a new, posthuman condition. First I will address the problem that our values may change when radical vulnerability transformations happen. Then I will respond to the epistemological problem of how we can know which transformations of our vulnerability and of our existential condition may take place in the future.

6.1.1 Dependent Values: A Proposal for a Modest Ethics

This subsection continues the discussion about values started in the previous chapter. Ethics understood as valuing is dependent on vulnerability transformations in several ways, which raises the following further questions.

First, the activity of valuing itself renders us vulnerable, as Nussbaum argued. To value something or somebody puts you at risk, makes you vulnerable. In that sense, ethics is not independent of the vulnerability transformations it is to evaluate. Is this a problem?

Second, values arise from vulnerabilities. Within the context of a certain practice, habit, action, or event, I become aware of a specific vulnerability (and respond emotionally to that awareness) and start valuing what is still not violated, not hurt, and not destroyed. For example, loving someone makes me dependent and vulnerable, and once I become aware of that risk, I fear that the loved one will leave me, will no longer love me, etc. But this also means that I really value that love relationship and that other. Another example: when I become ill, I become aware of my bodily vulnerability and

M. Coeckelbergh, *Human Being @ Risk: Enhancement, Technology, and the Evaluation of Vulnerability Transformations*, Philosophy of Engineering and Technology 12,
DOI 10.1007/978-94-007-6025-7_6, © Springer Science+Business Media Dordrecht 2013

then really come to value health. But if values are dependent and unstable in this way, how can they be independent criteria by which to judge vulnerability?

Third, to try to avoid the former problems, we may want to focus on judging *future* vulnerability transformations (in order to avoid the first problem) from a *macro*-perspective (in order to avoid the second problem). We take distance from the present and from the personal. However, even if we succeed in taking such a dissociative stance, our ethical outlook may change with future political, social, and other transformations. What is ethical valuing? In the universalist tradition of moral philosophy, if I say 'X is good' then I mean: 'This is not only good for me but good for all humans'. However, our view of the good changes and what we are as humans changes. Acknowledging moral change need not imply extreme moral relativism, since we can use concepts, values, and principles that have remained relatively stable during recent history. This is what we usually do in ethics: we do not radically question our concepts, values, and principles; rather, we apply them. We could also do this with regard to the evaluation of future technologies. However, we can only resort to this method on the condition that the world and what we are as humans—on which these concepts, values, and principles are based—also remains relatively stable. We can only use quasi-independent values if our condition remains similar. Our values are dependent on (a certain mode of) 'human being', which includes, and is partly constituted by, specific vulnerability patterns and configurations. A major problem with developing an ethics of future technology, therefore, is that technological changes may be so radical that they fundamentally change our relation to the world and therefore our vulnerability. How can an ethics of technology cope with this problem? Once again we must ask the question: how is a posthuman ethics possible?

Some believe there is an independent criterion to judge vulnerability transformations, for instance authenticity. Consider changes within a person's lifetime. Habermas appeals to 'being oneself' as an argument against the idea that someone else (my 'parents') programmes me genetically (before birth) and defines the capacity of being oneself as requiring 'that the person be at home, so to speak, in her own body' (Habermas 2003, p. 57). (Note that in itself this is not a good argument against genetic engineering, since the metaphor also allows for *rebuilding* your house (your body) or changing its interior to feel more at home.) Authenticity understood as 'being at home', however, is an instable and dependent criterion: it depends on the individual, on the moment in that individual's lifespan, and on the point of time in history. What it is to feel at home (in your body or elsewhere) and to be yourself may change radically in the future. Today we may feel that our existence is inauthentic and alienated, or would become so if a particular (enhancement) technology is implemented, but this says little about how we might feel in the distant future.

6.1.2 Imagining Future Vulnerability Transformations Under Conditions of Epistemic Opacity

In this section, I offer a heuristic to cope with the former problem. The epistemological problem is this: we do not know how the future world will look, how human being will change and what it will mean, and how we will relate to that future world. Without such knowledge, it is impossible to define our future vulnerabilities and values dependent on those vulnerabilities. How can we solve this problem, or at least make *some* progress in this direction? Are we entirely blind with regard to the future?

Note first that the epistemic problem an ethics of technology faces is not limited to the ethics of *future* technology. For example, as I argued before (Coeckelbergh 2010), there is already a situation of 'epistemic opacity' when we try to ascribe responsibility today. First, between our actions and the consequences of our actions lies a complex world of relationships, people, things, time, and space. Moreover, it is hard to distinguish between our contribution and those of others and between our actions and 'accident' or 'luck'. Thus, even in the *present,* we are often groping in the dark. The problem of evaluating future changes only adds to these existing problems.

How can we begin to close these epistemic gaps? My response to the problem of future vulnerability transformations is inspired by the work of Jonas and Anders (Jonas 1979; Anders 1956), who proposed that we rely on our emotional and imaginative capacities. Imagination not only helps us to imagine the current world, it also helps us to imagine the future. Both transhumanists and their opponents rely on speculative imagination: they explicitly and more often inexplicitly construct and assume future scenarios. Of course it is difficult to know which projections are 'realistic'. Habermas distinguishes between 'manifestations of a feverish imagination', 'serious predictions', 'displaced eschatological needs', and 'science-fiction science' (Habermas 2003, p. 42). But the lines between these notions cannot easily be drawn. Imagination is also at work when scientists project their 'serious' visions in research proposals and media interviews. And good science fiction, 'feverish' and eschatologically driven or not, helps us to explore moral questions regarding good technology.

Furthermore, proposing a 'heuristic of fear' (Jonas 1979) Jonas has argued that emotions aid us in defining what we really care about, what is at stake, and what we really value. Feeling teaches us what is at stake. Since it is easier for us to see the bad than to see that good, we must consult our fear(s) to understand what we really value. Jonas calls this imaginative-emotional effort (or, in his words, 'thought experiment') a duty (Jonas 1979, p. 64). He also refers to 'the serious side' of science fiction, which can help us in our heuristic exercise. For example, he mentions Huxley's *Brave New World*. Imagination, for Jonas, is not a private fantasy but a moral duty. Fear too is a preliminary duty (Jonas 1979, p. 392). He asks us to employ the vision of our imagination and our emotional sensitivity: try to imagine what happens to future generations, and let yourself be affected by it (Jonas 1979, p. 65).

About 20 years before Jonas wrote this, a similar view has been proposed by Günter Anders in *Die Antiquiertheit des Menschen* (Anders 1956). His way of defining the problem is particularly interesting since he does not see technology as something external and alien, as some philosophers of technology do, but as something that is deeply related to our (human) actions. He observed a gap between, on the one hand, what we can *do* with technology on the one hand (making things, *Herstellen*) and, on the other hand, what we can *imagine* (*Vorstellen*). We fail to emotionally and cognitively cope with what we make and their consequences (Anders 1956, p. 273). Anders proposes moral imagination to bridge this gap. He is aware of the limits of our imagination and feeling. For example, it may not be possible voluntarily to expand them. But as a moral person, we must at least will to try to break through the limits of imagination (Anders 1956, p. 273). Anders proposes 'moral stretch exercises' (*moralische Streckübungen*). This is a technique similar to religious and mystical techniques of self-transformation (*Selbstverwandlungs-Techniken*): we try to reach a region which we cannot normally enter (Anders 1956, p. 275). But here we try to reach artefacts, things we made ourselves but from which we are now experientially divorced.

Following Jonas's and Anders's advice, we could try to imagine future vulnerabilities and vulnerability transformations, including their material aspects and then evaluate them. But evaluate by which criteria? We may well use the values discovered with the aid of our emotions (in particular fear but also hope), but these reflect our *current* sensitivities. How will we emotionally respond to radically changed vulnerabilities and risks in the future? Can we anticipate these changes? And can we really imagine *radical* changes in our relation to the world, *radical* vulnerability changes?

Thus, there are at least four kinds of limitations a posthuman ethics faces. First, our imaginative and emotional capacities themselves are limited (a 'hardware' problem, if you wish). Second, the 'input' needed for this emotional and imaginative apparatus is incomplete (the problem of epistemic opacity). Third, in the future our vulnerabilities, and therefore what we will be, may change in ways which we cannot foresee. Finally, our values will change in ways we cannot foresee.

The issue is even more problematic since we know that even our more 'abstract' moral and other thinking (as opposed to the more 'lively' imagination) greatly depends on metaphors and that these metaphors depend on our *bodily experience* (Johnson 1993; Lakoff and Johnson 1980). If transhumanist human enhancement were to radically change our bodily experience, then our metaphors, and therefore our (moral) thinking, would change as well. Both our imaginative and our emotional capacities are tied to our current bodily and psychological make-up. We employ capacities to reflect on a condition that may involve a fundamental change in those capacities.

Furthermore, our current criteria for human flourishing, our human virtues, etc. are based on our current understanding of human being. It is likely that when human being changes, our ideal of what constitutes the good life and our ideal of the human will change as well.

I do not know how to solve this problem. However, I see the following possibilities of making some progress in this matter.

First, we can try to get as far as we can by means of some kind of 'moral stretch exercises', as proposed by Anders. Here the challenge is to find means to facilitate such exercises. What kind of 'sport' infrastructure should we build for the training of moral and epistemic imagination, and what kind of institutions does it need? For example, how can current information and communication technologies help? What kind of organisational structures do we need to train this kind of imagination?

Second, using imagination alone may be a much too detached way of exploring future possibilities. If we merely try to *represent* the future, we have only limited experience of what concrete technologies might do to us. We need more concrete experience, hands-on experience. We need to try out things and experience how it feels to handle them. We need know-how next to know-that. Of course we do not have these future technologies yet. They are literally not 'at hand'. We cannot perform with these technologies. However, artists and scientists may work together to give us *some* feel for how future technologies could reshape our experience. Although we can never gain complete know-how of the future, we can at least try to explore some possibilities.

Third, from a pragmatic point of view, it may be best to distribute our ethical attention in the following way: (1) some attention to the remote future and (2) most of our attention to the near future, which we are better able to imagine and cope with in a value-emotional way. We do not start from zero, we already have some knowledge. For that purpose, the history of technology and studies of current technology—what they do to human being and experience—can also help. What are the current frontiers of human vulnerability, and how are they stretched by emerging technology?

Fourth, we should acknowledge that ethics itself is *always* a hard exercise of balancing distance and proximity. Even if we are concerned with the present alone, we need to take distance from our situation in order to evaluate it, for instance by using ethical principles. At the same time, we want to be attentive to, and take into account, the particular features of the situation. Therefore, the uncomfortable and unstable situation the ethicist finds herself in when evaluating future technologies should not come as a surprise. We can only train more and try to achieve a better performance, a better balance between distance and proximity, and a better dance.

Fifth, it is important not to misunderstand the appeal to imagination made above. Some people wrongly think that imagination is epistemologically speaking free floating or arbitrary. This view may stem from a misguided interpretation of a correct analogy: the analogy between morality and art. The artistic imagination is not more 'free' than moral imagination: both are rooted in available knowledge and skills, in context, and in situation. Surely, imagination's epistemological anchor may not be as reliable as the scientific method is *supposed* to be, but it is not 'mere' fantasy.

Finally, the previous solutions seem to presuppose that there is a fundamental gap between humans and technology. But technology might also be part of the 'bridge' to knowledge of the future. I already said that technology can help us to explore future possibilities. But it does more than that: it also changes our experience of the world and our experience of the future. In this respect, it is worth noting that we already use *technology*—some would say: we *extend* ourselves with technology (see the next section)—in order to enlarge our experience and imagination. This is a kind of paradox and perhaps also a vicious circle or spiral: we use technology to solve a problem created by technology, but this creates a new problem, so we use new technology, and so on. For example, we can process many data in order to try to predict the future (e.g. the financial future), but in the end no one understands what is going on, and we become even more reliable on (new, 'better') technological systems. Or to give another example, now one that concerns the spatial gap: Internet-related technology makes a globalised world possible, but this globalisation also creates an epistemic gap. In order to cope with this gap, we use Internet, which again creates epistemic problems. And in the future, we might use a new kind of Internet and new technologies in order to try to solve the problem. But just as new technologies create new vulnerabilities, new *epistemic technologies* create new epistemic problems (next to other problems).

Using the insights from this section and the previous chapters, let us now return to the discussion about human enhancement.

6.1.3 Visions of Extensions and Enhancements: Mysticism, Violence, and Science Fiction

In the previous section, I proposed a heuristic of imagination and emotion: we should try to stretch our imaginative and emotional capacities in order to envisage and evaluate future vulnerability transformations, informed by the knowledge and narratives available to us. I also suggested that we already 'extend' ourselves—including our imaginative and experiential capacities—by means of technology. Let me illustrate this approach with regard to two technological changes: information technology (present but also the future) and advanced genetic engineering (remote future). This also helps me to further illustrate other dimensions of the ethics of vulnerability as an *imaginative* ethics: the first discussion will rely on thought experiments; the second will use literary fiction.

6.1.3.1 Human Enhancement by Information and Communication Technology as 'Extension'

One way of making oneself less vulnerable, it seems, is to 'extend oneself'. By using technology, one tries to overcome the limitations of one's human mind and body. This seems a key aspect of (transhumanist) proposals for human *enhancement*. Although the language of extension is still post-Cartesian and hence problematic (it still presupposes that there is something to be extended), let me first use the concept and then offer my objections.

Let me start with the claim that information and communication technologies (ICTs) extend us. If the capacities of our mind are limited, it seems reasonable to try to extend it with computers, electronic networks, mobile phones, and other electronic information and communication tools. Indeed, it seems that this is exactly what we are doing these days: we get used to thinking with electronic devices, that is, we think *through* electronic devices. We 'enhance' our thinking, our actions, and our experience by connecting to the Internet. The devices by which we 'connect' (almost) become part of us. The electronic device becomes a cognitive-perceptual prosthesis: it is an extension by means of an artificial device. In Ihde's words, we enter in an 'embodiment relation' with the technology. It is a kind of 'existential technological relation' because I perceive 'through such technologies' (Ihde 1990, p. 72).

Indeed, the new electronic technologies are our new glasses or telescopes: they mediate our relation to the world. Like glasses, they become 'part of the way I ordinarily experience my surroundings' (p. 73).

And just as the telescope bridges the epistemic gap (in time and space) between us and the stars, the Internet bridges all kinds of epistemic gaps. It is well on its way to be perceived as the 'universal' telescope; it seems that it claims to bring *anything* close to our eyes, in past and present, here and anywhere on earth and beyond. It is a platform for Google Earth and Google Street View, but it becomes a platform for 'Google Everything'. It appears to remove any distance. Extending ourselves with networks and search engines, we become nodes in the network and we become search engines ourselves. We become information and communication terminals. We become mobile phones. We become the screen on which the whole world can light up.

However, extending yourself does render you vulnerable: sometimes *more* vulnerable and in any case vulnerable in a different way. The transformations discussed earlier can be put in the language of extension: there are religious, social-political, psychological, and technological extensions—and therefore also different types of vulnerability transformations. In Chap. 4, I already used the language of extension in connection to ICTs: I suggested that by extending ourselves by electronic means such as the Internet, we make ourselves vulnerable in a different way. Indeed, using the new information and communication technologies generates new vulnerabilities: new kinds of viruses can invade us, we are not sure if we can trust the information that comes to us, and the possibility of social violence is extended beyond the circle of people we have direct face-to-face contact with or with whom we communicated by means of the 'old', offline technologies.

One way to respond to the problem that such extensions reduce but at the same time also increase or at least transform vulnerability is to further extend ourselves. We can use new tools in order to try to protect ourselves. But as I argued, this creates further vulnerability transformations and is likely to increase vulnerability (the vicious spiral of extension I mentioned before). It does not solve the problem.

Another way would be to try to *reduce* extension in order to become less vulnerable. But what does this mean? What is the opposite of extension? Does becoming less 'extended' mean retreating to some kind of 'inner self'? But the concept of authenticity understood in this sense is problematic: what are we *without* technology, and what are we without the world? Either we retreat to the self in the sense that we no longer act in and experience the world. This amounts to death. Or we relate to our environment, but it seems that without any techno-material interface between ourselves and our environment, we would find it very hard indeed to live our lives. Most non-human animals can do it, perhaps, but humans—as technological beings 'by nature'—cannot. We are the most *open* and hence the most vulnerable being. We are the being that stands most in need of extension and mediation by technological and social tools.

Can mysticism, in particular the kind of mysticism that strives for *oneness*, solve the problem? One could argue that we are too much dispersed and fragmented, rather than one, and that we should we try to become 'one'. To become 'unextended' then means to become one. But what does 'becoming one' mean here? On the one hand, if it means that we become one-as-separate-from-everything, then we have the same problem as mentioned above: this is not a human possibility and not a possibility for *any* living being. We are relational by nature. On the other hand, if it means that we become one with *everything*, then surely we are absolutely vulnerable. (Maybe this is happening today, at least if McLuhan was right that we are globally extended, but then it is happening *because of* information technology and not *without* technology.)

Finally, becoming 'unextended' may amount to the following idea, which is related to the idea of becoming one-as-separate-from-everything: we become one with a god who is not part of that 'everything' and is invulnerable—indeed share in the god's invulnerability or become an invulnerable god ourselves. Then we are so 'unextended' and invulnerable that we are no longer 'of this world'. But this is a very problematic 'solution', if *conceivable* at all. The cost of this kind of invulnerability, if such a state can be reached at all, is the end of human existence and life as we know it—involving again some kind of death. It would un-world us, since absolute invulnerability can only be reached if we leave the world altogether. Shedding our current vulnerabilities means to stop being human altogether; shedding *all* vulnerabilities means to stop being a worldly being. It is very doubtful that it is a real possibility, let alone that it is a *desirable* possibility.

However, should we use the concept of 'extension' at all? Let me pause and ask how helpful this concept is for my purposes. Is the Internet really an extension of our brain, of our body, and of us? What does this mean? First, it is important to be clear about what is extended. It seems that the concept of extension assumes that there is a 'something' (e.g. the mind) that is extended. This seems to assume a dualistic view that is not unlike that of Descartes: there is 'mind' and there is 'world'. But I have argued that being vulnerable is always a being-in-the-world, or presupposes being-in-the-world. Second, what is extension? What is the relation between technology and the human? The idea of technology as an extension (from Kapp to McLuhan and beyond) is attractive. But how should we understand that concept? The notion of extension can have a projectionist meaning (as in the work of Ernst Kapp), which implies that we *view* technology as an extension of ourselves, but this neglects what technology *does* to us, how technology shapes us, and how we in turn change the technology. The point is that technological artefacts are not just the screens on which we project our thoughts, but the things with which we inter*act* and that change and shape us. The term extension may mislead us to think that the object that is extended (the body, the self, the human) is itself something fixed. But this does not make sense. Technology changes our bodies, ourselves, and our ways of being. To use the term 'extension', therefore, tends to assume another dualism: between humans and technology. But our vulnerability is always already transformed by technology, and new technologies amount to new vulnerability transformations.

The concept of existential vulnerability I proposed avoids strong technology-body, technology-self, technology-human, and body-mind dualisms and the problems these dualisms create. Vulnerability does not start with a substance or essence that then gets extended (and potentially violated), but rather understands human nature as human *being*, with being meaning 'being-in-the-world' and 'being-at-risk'. We are not first 'natural' then 'artificially' extended. Rather, we are *already* and always 'extended': being at risk presupposes that we *already* relate to the world in various ways, which renders us vulnerable. Therefore, if we want to use the term 'extension' at all, we have to keep this in mind. Existence is already *ek-sistenz*, to use a Heideggerian neologism. We are already standing out to the world. And this standing out—in its active/controlling and emotional/receptive aspects—shapes our vulnerability. We are stand-up beings, who are always already on stage, exposing ourselves to the violence of the gods, the others, the technology, and finally ourselves. We are already naked in various ways. Our skin 'extends' to the world; we are 'spread out' in the world. (If we must talk in absolutes at all, we are rather absolutely vulnerable than absolutely invulnerable. But we can avoid this term 'absolute'.)

This vulnerability can only be undone by a process of de-worlding, which amounts to a process of dehumanisation and dying. To return to my earlier discussion of mysticism, we can try to conceive of the idea of 'overcoming' vulnerability by becoming *one* with the gods, with others, with technology, and with ourselves. Then we become one with what threatens us, what could violate us. As a result, we do not become absolutely vulnerable but absolutely invulnerable. If this ever happened, then our humanity would also be annihilated. We can only exist *as humans* by being resisted and by being in separation. At the same time, we are related to, and dependent on, the 'other'. There is always the possibility that 'that which resists me', that from which I am separated, violates me. We are never completely one with our environment; there is always an other (or, according to some: an Other). This is our condition of existential vulnerability and risk.

Of course, there is also the possibility that the divine other (the Other), the social other, the material other, or the self-other does *not* violate me. There are other possibilities of relating to the world, such as love. Violence is not the only possibility that emerges from the existential relation characterised by vulnerability (note that the possibility of love presupposes separation and vulnerability), and to describe our existential condition, we need more concepts than risk and vulnerability alone. In any case, the concept of existential vulnerability seems to allow a more adequate and broader view on human-technology relations and indeed human-environment relations.

Having said this, let me continue my discussion of how to evaluate new developments in ICT. Let us take a brief look at concrete human-computer interaction. Computers are not merely tools. They do things to us: to our work, to our lives, and to our identity. As Stahl remarks, the interaction between humans and computers is not value neutral, but changes the way we perceive ourselves and the way we can discharge our responsibilities (Stahl 2002). However, Stahl fears that if we see humans as machines, our sense of vulnerability may be lost. But this fear is unjustified.

First, leaving aside the normative question if we should see humans as machines at all (I think this is only *one* possible way of seeing humans and a very problematic one), it is important to realise that machines and technological systems are also vulnerable, only in a different way than humans. Machines do not have an existential vulnerability, but an 'essential' or 'objective' vulnerability. They are vulnerable in virtue of what they are and what they are is fixed. However, machines are not necessarily 'mere machines'. They can appear to us in various ways. What they 'are'—and therefore their vulnerability—cannot be defined apart from how they appear to us and apart from what they *do*; and what they do cannot be reduced to function, that is, to what they were *intended* to do. What technologies (particular tool, artefacts) do (see also Verbeek 2005) depends on the *inter*actions, the context, the constellation, and the situation it is part of. Technology is part of nature and culture; it is related to the world in different ways. But in contrast to humans, it is not aware of these relations, and it is not self-aware: it *is* related to the world, but does not consciously and actively *relate* to the world. In other words, technological artefacts do not have existential being like humans.

Second, assuming that it is bad for humans to become machines (in the sense of losing *existence*), when do humans become 'machines'? Where do we draw the line?

Let us start from the observation that we really care about drawing a line, about separating ourselves from technology. As I wrote previously, a negative anthropology needs technology and other things in order to define the human. For example, we define the human as not-robot (or as more-than-robot). And when it comes to risk, we talk about 'technological risk' as if it something entirely external. Technology is seen as something that stands 'outside' us and that produces the possibility of violence: in other words, technology makes us vulnerable. Although I noted that some separation must be presupposed in order to make sense of the human at all, I also questioned the view that technology is entirely external to us. We are always already related to technology.

However, we need not deny that emphasising the separation plays a significant role in shaping *the specific ways in which we relate to technology.* For example, it plays a key role in how we cope with vulnerability in the information age. As my remarks in Chap. 4 with regard to ICT already suggested, we could employ the term 'immunity' (e.g. as borrowed from Sloterdijk) to describe the anti-vulnerability strategy involved when we reluctantly open ourselves up to, and extend ourselves to, the Internet and other digital information and communication technology. We try to gain immunity: we try to protect ourselves from electronic viruses, from violations of our privacy, etc. by using virus scanners and firewalls. But this strategy is not entirely peaceful: we have to destroy the intruder. The analogy made here is between biological immune systems and electronic immune systems, which in turn can be described in analogy with Warcraft: we try to detect the enemy on time and destroy him before further harm can be done.

One thing we can learn from such an anti-vulnerability strategy ('strategy' also fits the military analogy) is that by trying to become less vulnerable, we also produce violence. Many military 'defensive' actions produce violence. Biological and informational immunity warriors produce violence. We already learned that anti-vulnerability strategies transform our vulnerability, but now we arrive at the insight that such a transformation can involve *violence against that which threatens us*. In the case of electronic viruses, this may not pose an ethical problem, but if we consider threats by entire computer systems on which other people depend and if we consider (low tech) threats by others (other human beings), then the following ethical question arises: when is it justified to use violence in order to reduce your own vulnerability? What means are justified to reduce your (own) vulnerability? We need ethical constraints on violent anti-vulnerability strategies. We need peaceful ways of coping with our vulnerability, which lies between two opposite poles. On the one hand, the mystic may try to become one with the world but tries to do that 'in her mind', that is, she searches for spiritual unity. This seems peaceful, but does not *necessarily* exclude violence towards others and towards the world—in fact against anything that threatens the peace of mind or the community of mystics. On the other hand, someone may seek to reduce her vulnerability by trying to literally control the whole world, by becoming a god-like dictator. If you control everything, there is no longer a threat from 'outside', herein lies the temptation. But violence is the result.

Therefore, it seems that we better protect ourselves in a more passive way, which does not attack the attacker. But is this solution feasible? How peaceful can we make it? As I said, in case of an attack, I might 'have' to use violence to protect myself. Moreover, the very attempt to control others in order to reduce your own vulnerability and that of your society is common and should not strike us as alien to the history of human culture and civilisation. Politicians have always tried to justify offensive wars as a defence, as a vulnerability-reducing means. Citizens have always justified street violence in terms of defence. And attempts to control our natural environment have been violent as well. We 'defend' ourselves against 'nature' by 'fighting back'. We violently transform our environment and kill any animal we perceive as threatening us or our activities (e.g. agricultural activities). (Note again the process of externalisation: the environment appears as an external 'nature'.) By themselves, these observations do not justify the violence, but the question is to what extent we can avoid some form and some degree of violence as relational beings. The question of degree is important here: efforts to control others and the world need not strive for *total* control, and we could also try to minimise the violence involved in our anti-vulnerability actions. Maybe this is *the* ethical question with regard to anti-vulnerability actions: how to minimise both vulnerability and (the possibility of) violence. This means how to simultaneously minimise *our* vulnerability and the vulnerability of *others*.

Thus, we can conclude that our struggle against vulnerability is not existential-ethically neutral for two reasons: (1) since it transforms our own vulnerability and (2) since it *may* be a source of violence against the other—human or non-human—who/which threatens us. In the end, it may increase the vulnerability of that other. This observation is politically relevant as well: vulnerability struggles are, or have at least to potential to become, power struggles—between gods and humans, between humans, between humans and natural beings, between humans and technology, and within ourselves. This conclusion, of course, does not preclude the possibility of more harmonious relations with others and with the world (as an alternative to or as another side of power struggle)—even if that other renders us vulnerable. As I argued before, vulnerability itself is not only something bad but also a necessary condition for what we value most in life. Moreover, inaction is not an option: we *have* to cope with our vulnerability. The ethical question is not *if* we should do something to others and to our environment in response to threats to our vulnerability (thereby transforming both ourselves and the not-self), but what we should do and *how* we should do it.

Therefore, ethics of technology understood as a normative anthropology and as an ethics of vulnerability should critically examine what specific technologies do to our vulnerability and how we cope with our vulnerability in response to each technology.

For example, do new social media increase vulnerability? In one sense, they do: people expose their private life to others, which renders them vulnerable to misuse and abuse of these data. This constitutes a kind of 'social vulnerability' and 'data vulnerability' (again the technological and the human are intertwined). In another sense, one may ask if people really *appear* vulnerable on the Internet. Social media seem to preclude the possibility of true friendship, which requires a setting in which the friends disclose each other, that is, where they can appear vulnerable to one another. Perhaps such disclosure was crucial to the traditional humanist idea of friendship, which—as Sloterdijk has reminded us (Sloterdijk 1999)—was fostered by letter writing (and face-to-face contacts, of course). On the Internet, by contrast, people seem to present only a very selective image of themselves, meant to hide their vulnerability rather than expose it. As moderns who value invulnerability, we learned to present ourselves as invulnerable.

However, as Briggle suggests in his insightful article on friendship and the Internet (Briggle 2008), the Internet *does* provide possibilities for friendship as self-disclosure, self-exploration, and self-evaluation in writing, which I guess modern humanism (coupled with Romanticism) requires. If this is true, such practices do not necessarily increase or decrease the vulnerability of the self, but at most transform it. It is transformed, since the medium and the culture will have *some* influence on the message. For instance, if Internet practice is embedded in a culture of speed (as Briggle and others argue, see Briggle 2008, p. 78) and if it may be less able to guarantee privacy,[1] or if it uses a format such as blogs where exposure is the purpose, then vulnerability changes and so does the culture of humanism continued or echoed by such practices.

6.1.3.2 Human Enhancement by Genetic Engineering and the Cost of Emotional Invulnerability

The usual ethical question with regard to human enhancement by genetic engineering is as follows: are we morally allowed to intervene in the human genome? This question creates various (further) problems. For example, human enhancement by genetic engineering is usually distinguished from medical or therapeutical genetic interventions. However, the ethical border between the two is not very clear. For instance, genetic interventions that aim to prevent diseases such as cancer and Alzheimer's are likely to prolong our lives as well. Furthermore, we must ask how relevant the time of intervention is: does it matter, ethically, if we intervene prenatal rather than postnatal? I have already touched upon these or similar questions in

[1] Note that the old mail system was not perfectly safe either.

the second chapter, and it is important to discuss them with regard to present or near-future ethical and political decision-making.

However, we can also ask other, perhaps more important questions: what aim(s) we want to reach with these technologies? In particular, we must also ask philosophical-anthropological and social-philosophical questions: what kind of being do we want to be, and why? And what kind of society do we want? How do we want to relate to one another? This requires us to further reflect on future possibilities—including the more remote future. As I have argued before, imagination can help us with this task. What can imagination do for the discussion about human enhancement and vulnerability?

Let me show how we can use literary fiction for ethical reflection on human enhancement and posthuman vulnerability by interpreting *La Possibilité d'une île* (2005) by Michel Houellebecq. This novel aids our moral imagination, in the sense that it helps us to explore the ethical and social-philosophical problems connected with a certain kind of genetic (and social) engineering programme. Fiction does not only help us to explore new technological possibilities, but (therefore) also new moral, anthropological, and social possibilities.

In his novel, Houellebecq presents two contrasting images of human being and society. Contemporary humans suffer from a lack of love, struggle with their sexual desire and emotions, are troubled by ageing, and are unable to live with others in a decent way. They live in a society in which youth, sex, and pleasure are emphasised. This renders them vulnerable and prone to suffering, since their desires are continuously stimulated. They are frustrated. Houellebecq's future humans (or posthumans), by contrast, lack strong desire and emotions, live forever in the sense that they are cloned and live in nearly absolute isolation: separated from each other and separated from the 'wild' humans who are still emotional and primitive. In this way, their emotional and their socio-psychological vulnerabilities are diminished.

This picture can be seen as a response to two philosophical questions, one about the (individual) good life and one about how to live together.[2]

As for the good life, there always has been a tension between hedonist and anti-hedonist views. For Houellebecq, the first view is embodied in our contemporary consumer society, a world of eternal *kids*, which stimulates desire to unbearable levels but makes their realisation impossible. The author's descriptions of human suffering are very similar to those of Hindu and Buddhist teaching. The human (Daniel 1) only knows short moments of pleasure and extremely long periods of frustration. The posthumans (clones Daniel 24 and 25), however, are beyond this kind of suffering. Genetically engineered to be calm and contemplative, they live the life many philosophers and religious people have always strived for. However, Houellebecq's vision differs from traditional anti-hedonist strategies. First, it is deterministic (it is suggested that humans are entirely predictable), which runs against the view of many traditional philosophers and theologians, who wanted to

[2] Of course the two questions are related—we are social beings and human good cannot exist outside the social—but a logical distinction can be made for the purpose of this analysis.

preserve or even celebrated human freedom. Second, the method to reach the reasonable, contemplative life is no longer socialisation and training (disciplining, *askesis*) and/or isolation (retreat), but is mainly[3] realised by changing the human genetic code. It is a *technological* solution.

As for living together, the old questions if and how it is *possible* to live together for humans are answered negatively in the novel. Houellebecq emphasises a specific kind of social vulnerability: by showing a posthuman 'society' in which people live in isolation, he continues in the tradition of what is often interpreted as Sartrean thinking on this issue, the pessimistic view of the possibility of living together exemplified by the 'hell are the others' one-liner known from the theatre piece *Huis clos*. The message seems to be that it is better for us to live alone; in this way, we avoid the violence others do to us. To put it in Hegelian terms: if we live together, we can only objectify the other, and we are always objectified by others.

Indeed, when describing current modern society, Houellebecq depicts a *non-society*. In his pornographies of the nonsocial life, there are only interacting human bodies, a kind of zombies. Sometimes there is pleasure, but never joy. Elderly people are excluded from the lustful but soulless parties of the young. The posthumans, by contrast, do no longer aspire to social living. Physical contact between them is abolished. Everyone lives single in a gated community, which does not only exclude non-posthumans but also (at least physically) isolates the posthumans from each other—a community without community. Outside 'wild' humans engage in primitive violence; inside there is peace and calm. Fear is absent, including fear of death. There is still some contact between the posthumans via technological means, comparable to contemporary chatting, but this is the only contact the posthumans have and the temperature of the relations never rises above freezing point. It is also a radically unequal society: the 'wild' humans are, in fact, the *Untermenschen*, which are left to violence and suffering. They are radically and literally excluded.

Interestingly, the author does not remain content with a description of these two worlds (a contemporary one and a future one), but also offers an evaluation, a normative stance—at least if the reader does some work of interpretation. First, the book can be read as a critique of contemporary hedonistic and consumptive patterns, which increase socio-psychological vulnerability. Second, with its descriptions of psychological training, it gives *some* credit to the traditional *askesis*-alternative (albeit realised by technological means); in this sense, it can be read as a continuation (but also, modification) of the Western humanist tradition rather than its abolishment. Finally, this becomes especially apparent when the book suggests that always trying to *enhance* humans is the real misanthropy (a term opponents of Houellebecq use to discredit him); rather, we are warned that we may pay too high a price for human enhancement.

This conclusion emerges towards the end of the book. Some new humans are not happy with their lives: they want an intense life. In particular, they want love. But love cannot be disconnected from emotional suffering. Houellebecq describes love

---

[3] There are still some disciplining techniques, exercises which explicitly refer to the Buddhist tradition.

in terms of mutual oppression, torture, and death. Happiness is not on the horizon. But it is a *human* life. In such a life, joy and suffering are part of one package deal. If posthumans choose this human life, life becomes real. Humans are able to dream, to dream of community, to dream of love, to dream of endless happiness. Again (and here most clearly), it turns out that Houellebecq's novel is a praise of humanity rather than a plea for its abolishment.

This literary-philosophical exploration helps us to critically evaluate a particular vision of human enhancement, not in its abstraction and entirety, but in detail and in experience. If we take seriously the insight offered by Houellebecq's novel, we are better prepared to evaluate those proposals that attempt to change human emotional vulnerability. Do we really want to eliminate emotional suffering by technological means? What could we gain, and what would we loose on the way? What is worth preserving about human being? Houellebecq's novel suggests that the human vulnerability configuration is a complex whole and that if we diminish one vulnerability, we may increase other vulnerabilities and even remove a precondition for what we value most. Vulnerability has an ecological, holistic structure: it depends on many relations, and a change in any of its parts has consequences for the whole. Moreover, the relation of emotional vulnerability to the good is at least as ambiguous as emotional experience itself. One this point, Houellebecq's apparent antihumanism is in complete agreement with the humanism of Nussbaum and ethics of care theorists: vulnerability is not only something we necessarily need to get rid of, as some transhumanists aim for; by enabling us to live, aspire to, and dream of good and flourishing individual and communal lives, it is not only something to deplore; it is also a precondition for the emergence of the good.

6.1.4 Conclusions for the Discussion About Human Enhancement and Transhumanism

The transhumanist project of human enhancement must be understood as a collection of proposals that share the goal of a radical technological transformation that will alter our vulnerability patterns and therefore change what we are. In itself, technological transformations of vulnerability are not new; our vulnerability has already been changed in the course of human history and so have human beings. Moreover, changing ourselves—by technological and non-technological means—has always been part of the humanist project. The novelty, if any, lies in the extent of the change promoted by transhumanists and in the specific technological means proposed (as opposed to other technologies and other means). This raises serious ethical questions, since in these visions the definition and the future of human being is at stake. With regard to the issue of vulnerability, we should explore the precise ways in which transhumanist proposals may change our vulnerability and critically evaluate these changes. Ideally, we must use our imagination not only to project vulnerability changes in the near future but also to foresee how our values may change accordingly, since they are at least partly dependent upon vulnerability changes. We must try to stretch our imagination and our moral sensitivity, and we must explore future experiential possibilities by literally getting in touch with artistic and scientific experiments.

Of course, one could argue that since it is hard to foresee the remote future, we should concentrate most of our efforts on evaluating human enhancement in the near future (within 20 or 30 years). However, we should not neglect the more fundamental ethical, philosophical-anthropological, and social-political issues raised by the transhumanist vision, since it attends us important and difficult philosophical questions that do not only ask us to make explicit how we want to shape the techno-anthropological future (at least to the extent that this is possible) but that are also relevant for today. We should not be blind to on-going vulnerability transformations by technologies already available, in particular current information and communication technologies (hence the '@' sign in the title of this book). In order to be able to look at these changes from a distance, it can help to think about the remote future and discuss 'unrealistic' and 'science-fiction' visions of that future.

This conclusion is neither a 'yes' to the transhumanist project nor an *a priori* and absolute 'no'. Should we preserve human being as it is now? If the question is put in such absolute terms, it is hard to answer. Human being has many aspects, is changing in various ways, and is changing in relation to many developments. Is there a 'reason' for preserving humanity as such? Is there a 'reason' for keeping anything in existence? Unsurprisingly, Jonas goes into 'metaphysical' questions when discussing the future of humanity. But is this the way forward? To put it as a metaphysical problem is not wrong per se, of course, but the advantage of my response to the transhumanist challenge is that it sketches an approach that enables us to ask more precise and more concrete questions. If we understand 'humanity' as variable and as being constituted by various vulnerabilities, we can carefully study past changes and try to foresee future changes of specific vulnerabilities. Instead of evaluating 'humanity' as a fixed and single entity, we can evaluate specific past, current, and near-future vulnerability changes. If we ask 'should we preserve human being?' we ask—among other things—if we should preserve the current vulnerability patterns given the current technological and other developments. But these patterns and developments—albeit related—can be analysed separately, which allows a more precise discussion: we can ask how specific emerging technologies (may) alter our vulnerability. This approach does not exclude the possibility to bring metaphysical, religious,[4] anti-metaphysical, anti-religious, emotional, rational, and other perspectives to bear on the problem, but it provides a way to channel ethical effort away from the 'big' question to slightly smaller problems that are a little more within the reach of our limited cognitive capacities as human beings and that are practically relevant.

[4] The ethical-anthropological approach to human enhancement defended in this book is in line with what some religious thinkers do. For example, recently Michael Hogue has used insights of Jonas and Borgmann to formulate a 'biocultural theological anthropology'. He asks us to think about how technology is fundamentally transforming human moral life and reminds us of 'our ultimately limited capacities to control the world' (Hogue 2007, p. 92).

Moreover, in contrast to determinist versions of transhumanism (as e.g. 'Singularity' thinking), this approach acknowledges that we are not helpless with regard to existential vulnerability changes, but that we can—to *some* extent at least—influence what we are to become by trying to understand our current vulnerabilities, project possible future changes to these vulnerabilities, and evaluate these possibilities to our best abilities, that is, as well as we can within the limitations of our current imaginative, emotional, and rational capacities.

Apart from this focus on smaller problems (particular vulnerability transformations and their ethical implications), however, we can also formulate more general guiding principles for the discussion about human enhancement. Particular vulnerabilities and vulnerabilities transformations do not take place in isolation. If we consider again my discussion of Houellebecq's novel, the main lesson is perhaps that human being should be understood in a holistic way. It is risky to change one aspect, since that has also consequences for other aspects of our existence. In other words, human being can be described as an existential ecology, and therefore its vulnerability must also be understood holistically and ecologically. As we are about to learn from what happens to the ecology of our planet as a result of our uncontrolled, uncoordinated, and largely unregulated interventions, we should engage in a similar exercise with regard to human being. We are already changing human being by uncontrolled, uncoordinated, and largely unregulated interventions. Moreover, if technological changes are proposed to alter human being in a more radical and purposeful way ('more' since we have already changed), we should be aware that whatever 'human enhancement' intervention we do will affect other aspects of our vulnerability. Any change to a particular kind of vulnerability is a change to the ecology of vulnerabilities. For instance, Houellebecq's novel suggested that if we make human beings less emotionally sensitive in order to reduce suffering, we touch other aspects of what we are—for example, it would threaten our capacity to experience joy. Thus, the emotional psychology related to our vulnerability is a holistic ecology, and if we change one aspect, we change other aspects. Furthermore, human being as an ecology is also connected to other ecologies: social, cultural, political, and environmental ecologies. (I will discuss some of these connections in the next chapters.)

If this existential-ecological view of human being and human vulnerability makes sense, we have a good justification for taking precaution when trying to change human being, since our knowledge today is very limited and does not match at all the complexity of human being. We are only at the beginning of what should be a trans-disciplinary effort to develop an ecology of human being and existential vulnerability. So far, various disciplines have only contributed fragmentary knowledge of the connection between different vulnerabilities. For example, we still know very little about what is referred to (in modern thinking) as the relation between 'body' and 'mind'. A good degree of precaution, therefore, is advisable. And if we do intervene in the human genome and make the change inheritable, for example, if we disable a gene that causes a terrible disease or extend ourselves with a new life-changing ICT device, we should explore as far as we can, with science and imagination, the consequences for other parts and aspects of the vulnerable ecology that constitutes us.

To conclude, even if there is no *a priori* reason why we should not change human being (as I argued in Chap. 2), there are good reasons why we should closely monitor current technological developments that move in the direction of radical human enhancement. But should we also *stop* some of them? One the one hand, one could argue that one should halt some technologies given the overwhelming uncertainty about its impact on human being. In other words, as long as we do not know what we are doing, we should stop or slow down for a while. On the other hand, this measure is problematic if we consider the epistemological problem again. To really know what technologies might do to us in the future, we need to experiment. We need the sciences and the arts in order to explore future possibilities. We need imagination-in-action. Thus, to be cautious does not mean that we should only dwell in the land of contemplation, awaiting better times for science to finally invade the territory preyed upon. Rather, we should act and decide: develop policies and make political decisions. But in order to do this in a well-informed and a wise way, we must focus our attention on the active and on-going scientific, artistic, cultural, and philosophical exploration of the universe we wanted to travel beyond, but still do not know enough about that wondrous and vulnerable whole called human being. If we expand our knowledge in this way, we will be able to better imagine an existence and a condition that transcends ours in a desirable way. Perhaps such an existence will remain within the boundaries of what today we call 'humanity', perhaps not. But attention to the normative dimension remains crucial and cannot be disconnected from the descriptive effort of existential ethical-anthropological analysis. This requires science but also science fiction. If we are going to improve ourselves in a way that is more radical than ever before (if such radical changes are possible at all), we better know not only what we are doing but also what we want to become. Human as we are, this is still unclear in the twilight of the new existence promised by the transhumanist prophets.

6.2 Exploring Posthuman Vulnerabilities

In spite of the epistemological limitations to a 'posthuman' ethics and the need for imagination-in-action, we can still say something about future, 'posthuman' vulnerabilities by further developing what I said about past and present *human* vulnerability transformations and by drawing on science fiction. The following exploration and discussion of posthuman vulnerabilities supports my thesis that in so far as transhumanism aims at invulnerability, it must necessarily fail. If we cannot defeat the dragon, what would be our posthuman dragons, if we would ever reach a 'posthuman' stage at all? Let me draw on recent work (Coeckelbergh 2011) to list and discuss what I consider major sources of posthuman vulnerability.[5]

[5] A first version of this discussion has been published in the *Journal of Evolution & Technology* (JET).

6.2.1 Physical Vulnerabilities

Even if we were to (further) enhance ourselves by building more electronic, medical, and other walls, shields, and harnesses, we could never fully protect ourselves against physical threats. Other posthumans (or humans, if they would still be around at all) could harm us and there will always be external forces that are not within our control. Surely there would be new protections and immunities against specific threats. For example, posthumans might be immune for many diseases that kill humans today. However, we will never be able to be immune against all possible health threats. The history of medicine technology shows that for every disease new technology helps to prevent or cure, there is at least one new disease that escapes our techno-scientific control. We can win one battle, but we can never win the war. There will always be new diseases, new viruses, and, more generally, new threats to physical vulnerability. The military metaphor is not a coincidence: what we may call the 'dialectic' of anti-vulnerability struggle is typical for the development of military technologies. There are always new weapons but also new anti-weapons and new shields. Today new rocket shields are being developed together with new rockets, and current developments in military robotics will most likely lead to the development of 'anti-robot technology'. Moreover, 'old' risks may not disappear, although the specific vulnerability might be transformed. Consider natural disasters caused by floods, earthquakes, volcanic eruptions, and so on.

Furthermore, the very means to fight those threats sometimes create new threats themselves. Consider the use of antibiotics that leads to the development of more resistant bacteria or new security measures in airports which are meant as protections against physical harm by terrorism but which might pose new risks related to privacy and perhaps health. Paradoxically, technologies that are meant to reduce vulnerability often create new ones. This paradox of vulnerability is also applicable to posthuman technologies. For example, posthumans would also be vulnerable to at least some of the risks Bostrom calls 'existential risks' (Bostrom 2002), which could destroy 'posthumankind'. For example, nanotechnology or nuclear technology could be misused, a 'superintelligence' could take over and annihilate humankind, or technology could cause (further) environmental pollution, resource depletion, and ecological destruction. Military technologies are meant to protect us but they can become a threat and make us vulnerable in a new way. The atomic bomb is a good example of this paradox (or perhaps one could call it a dialectic—the dialectic of vulnerability). Moreover, technological development seems to threaten the conditions of living it was meant to improve by using up natural resources. We wanted to master nature in order to become less dependent on it, but now we risk destroying the ecology that sustains us. And of course there are many physical threats we cannot foresee—not even in the near future. Posthumans will remain vulnerable to at least some existing physical threats, but they will also face new risks and create new vulnerabilities.

6.2.2 Material and Immaterial Vulnerabilities

Physical vulnerability depends on the body. But this body is not a 'given'. We have always extended the human body with technology. This process of *cyborgisation* is likely to increase in posthumans, who would extend themselves with information technology and other technology to a much higher degree than contemporary humans. As I suggested before, if we are 'extending' ourselves with technology, then this has implications for our vulnerability. For instance, today we already 'extend' our body and mind by means of information technology. But that technology is also vulnerable to threats of various kinds. I already mentioned computer viruses. Codes can be accessed and changed (and increasingly, biological 'codes' can be changed too). Like biological viruses and weapons, fighting that risk involves on-going cycles of threats, counter-measures, and new threats. (We may also consider physical damage to computers and other hardware, although that is much less common.) If posthumans would also live in an 'infosphere' (see e.g. Floridi 2002) or cyberspace, then it is important to realise that this is not a sphere of immunity. Perhaps our vulnerability becomes less material, but we cannot escape it.

For instance, a virtual body in a virtual world may well be shielded from biological viruses, but is vulnerable to at least three kinds of threats. First, there are threats 'within' the virtual world itself. Consider for instance the threat of virtual rape. Such threats constitute what we may call 'virtual' vulnerability, which is similar and dissimilar to 'real' vulnerability. It is dissimilar since there is no 'physical' threat, but it is similar since the psychological harm is similar: *the experience, the habit, and the practice* are similar in some ways. (For example, I have argued that this is the moral problem with violent computer games: the violence is not real, but the habit is similar, which is relevant for virtue ethics (Coeckelbergh 2007).) Second, the software programme that provides a platform for the virtual world might be damaged, for example, by means of a cyber attack. Damage to the code can lead to the 'death' of the virtual character or entity. Third, all these processes depend on (material) hardware. The World Wide Web and its wired and wireless communications rest on material infrastructures without which the web would be impossible. Any 'second' world depends on the 'first' world. Therefore, if posthumans uploaded themselves into an infosphere and dispensed with their biological body, they would not gain invulnerability and immortality but merely transform their vulnerability. They might become a kind of 'cyber ghosts', but their 'ghostly' existence would be highly vulnerable given these and other immaterial and material threats.

6.2.3 Bodily Vulnerabilities

We might fantasise about immaterial and invulnerable existence in the infosphere, but seldom have these fantasies dispensed with a body (although not necessarily a human body). One reason is probably because as far as we know, we *need* one. Minds need bodies. Contemporary research in cognitive science establishes that embodiment is necessary: it is argued that minds can only develop and function in interaction with their environment (Lakoff and Johnson 1999 and others). This is also acknowledged by most transhumanists. In his famous 1988 book on 'mind children', Moravec already argued that true AI can only be achieved if machines have a body (Moravec 1988). And Kurzweil writes:

> If we are truly capturing a particular person's mental processes, then the reinstantiated mind will need a body, since so much of our thinking is directed toward physical needs and desires. (…) The human body version 2.0 will include virtual bodies in completely realistic virtual environments, nanotechnology-based physical bodies, and more. (Kurzweil 2005, p. 199)

Thus, if posthuman existence is still somewhat human in terms of the kind of thinking involved (and it seems most transhumanist would like to keep the basic way we think), then uploading and nano-based cyborgisation would not dispense with the body but transform it into a virtual body, nano-body, or another (semi-)artificial or cyborg body. This would necessarily create vulnerabilities: it would create vulnerabilities that resemble the vulnerabilities we know today (for instance virtual violence), but it will also create new vulnerabilities specific to the new body. For instance, no one knows what kinds of risks would emerge when we had nanorobots in our bloodstream. Our bodies would be transformed in always that are hard to imagine, and so would our vulnerability.

6.2.4 Metaphysical Vulnerabilities

Posthumans would also remain mortal in the following sense. According to an influential metaphysical doctrine, bodies are organisations of matter, in particular organisations of elementary particles. The particular combinations of matter are always temporal since vulnerable to disintegration. The Greek philosopher Democritos, known as the founder of atomism, already claimed that whereas atoms are eternal, the objects composed of these atoms are not. Worlds come and disappear again. And while contemporary physics and metaphysics are no longer atomist in the common sense of the word ('atoms' turned out not to be the smallest particle), physics is still after elementary particles and the natural sciences embrace an atomist metaphysics concerning the relation between systems (or organisms) and their elements. Even the infosphere has its 'information objects', which might be interpreted as compositions of 'elementary particles': bits.

With this atomism comes an atomist view of death: there is always the possibility of disintegration; neither physical-material objects nor information objects exist forever. Information can disintegrate (when the elements no longer combine to form that particular order) and the material conditions for information are vulnerable to disintegration as well. Thus, at a fundamental level, everything is vulnerable to disintegration, understood by atomism as a reorganisation of elementary particles. This 'metaphysical' vulnerability is unavoidable for posthumans, whatever the status of their elementary particles and the organs and systems constitutes by these particles (biological or not). According to their own metaphysics, the cyborgs and 'inforgs' (Floridi 2008) transhumanists and their supporters wish to create would be only temporal orders that have only temporal stability—if they have any at all. Elementary particles may be 'immortal' in some sense, but we are not.

Recently both philosophy of information (in particular Floridi's work) and contemporary physics seem to move towards a more ecological, holistic metaphysics, which sees the infosphere or the universe in relational terms. I believe this suggests a different definition of death. In information ecologies, perhaps death means the absence of relations, disconnection. Or it means: deletion, understood ecologically and holistically as the removal out of the whole. But in the light of this metaphysics too, there seems to be no reason why posthumans would be able to escape death in this metaphysical sense. Whether they are seen as composed of elementary particles or as relational nodes in a network-ecology, they remain vulnerable and 'mortal', however virtual they might have become and whatever that means for the *experience* of posthuman dying and posthuman disease.

6.2.5 Existential and Psychological Vulnerabilities

Vulnerability not only has its source in material-ontological reality, but also in existential experience, psychology, and perception. We are not only 'directly' vulnerable as bodily, material, and (meta)physical entities; as humans we can also *know* and *experience* those vulnerabilities—or rather we 'have' to: we can only experience vulnerability in consciousness, by becoming aware of it—for example, by imagining the possible violation. In this sense, there is no 'direct' vulnerability; vulnerability is always experienced. This is the subjective dimension of being vulnerable, of being at risk I introduced in Chap. 3. It also involves what I have called 'second order' vulnerabilities. For instance, we can become aware of the possibility of disintegration and the possibility of death. We can also become aware of less threatening risks, such as disease. There are many first-order vulnerabilities of which we can become aware. Awareness of them renders us extra vulnerable as opposed to beings that lack such an ability to take distance from ourselves. From an existential-phenomenological point of view, but also from the point of view of common sense psychology, we must extend the meaning of vulnerability to the sufferings of the 'mind'. Vulnerability awareness itself constitutes a higher-order vulnerability that is typical of humans. In posthumans, we could only erase this hyper-vulnerability if we are prepared to abandon the particular higher form of consciousness we 'enjoy' as humans and which is a condition of possibility for the conscious anticipation of death and its related sufferings. No transhumanist would seriously consider that solution to the problem; we like to keep our higher form of consciousness. Therefore, if posthumans were to have a higher form of consciousness not too dissimilar to ours, then they would have to cope with second-order vulnerabilities as well as first-order ones.

6.2.6 Social and Emotional Vulnerabilities

Sufferings of the mind concern emotions and others. I already argued that there is an important social dimension to human vulnerability. We do not live in isolation: we are social beings that depend on each other for fulfilling our physical, emotional, and other needs, and this makes us vulnerable in many ways. We tend to form relationships, groups, and communities, which leads to many advantages as well as plenty of possibilities for suffering and violence. More generally, if I depend on you socially and emotionally, then I am vulnerable to what you say or do. Unless posthumans were to live in complete isolation without any possibility of inter-posthuman communication, they would be as vulnerable as we are to the sufferings created by the social life, although their particular forms of sociality and the precise relation between their social life and their emotional make-up may differ.

This also recalls my analysis of Houellebecq's novel. We can imagine posthumans who would spend their lives in isolation, as hermits living in 'compounds' that are fenced off from the harsh natural world and from 'degenerated' humans who revert back to 'primitive' and violent forms of group life. However, we might also imagine that such posthumans would still feel the need to communicate with others. If they had still this need at all, then the tension we modern humans know between trying to reach immunity but being caught up in social-relational dependency, which is typical of modern humans who seek autonomy, would remain in place. If they had other values, these values would be related to the social as well. Only fiercely antihumanist enhancement would abolish social relations *and the need for social relations* entirely.

Of course even in a non-isolationist vision, posthumans may be changed in such a way that they have a different social and emotional life. But then these more radical changes would create new and different vulnerabilities. For example, as I have suggested in my interpretation of Houellebecq's novel, a type of emotional 'enhancement' that would diminish our desire would come at a price: such posthumans might have a reduced capacity to feel sad, but at the cost of a reduced capacity to desire and to feel joy. Again, we must ask ourselves: are we prepared to pay that price? Even if we succeed in diminishing this kind of vulnerability, we might lose something that is of value to us. This brings me to the next kind of vulnerability.

6.2.7 Ethical-Axiological Vulnerabilities

Humans are not just witnesses and interpreters of physical, bodily, psychological, and social processes. They also evaluate them and engage with them. But the very activity of *valuing* renders us vulnerable. We value not only people and our relationships with them; we are also attached to many other things in life. Caring makes us vulnerable (Nussbaum 1986—I already mentioned her view in the previous chapter). We develop ties out of our engagement with humans, animals, objects, buildings, landscapes, and many other things. This renders us vulnerable since it make us dependent on (what we may experience as) 'external' things and people. We sometimes are emotional about things since we care and since we value. We *suffer* since we depend on external things and people. Valuing is a source of joy but also of harm.

The Stoics knew this and followed a particular strategy of immunity: they tried to disarm the emotions and the vulnerability by not caring about the externalities, that is, by trying to cut the (emotional psychological) ties and the dependencies.[6] However, this Stoic solution—if implemented in an absolute way—is both impossible and undesirable. It is impossible since we can never completely detach ourselves from the world; this would be death. Moreover, it is undesirable since, as Nussbaum argued, if we remove our dependencies, we also remove the possibility for the good. Ultimately, therefore, the Stoic immunity therapy is not a solution to the problem of vulnerability.[7]

Surely, posthumans could be cognitively equipped to follow the Stoic strategy to some degree, for instance by means of emotional enhancement that allows more self-control and prevents us forming too strong ties to things. And if we really wanted to become invulnerable in this respect, we could create posthumans who no longer care at all about external things—including other posthumans. They would be posthumans indeed, since they would no longer have the ability to care and to value. They would 'connect' to others and to things, but they would not *really* engage with them, since that would render them vulnerable. They would be perfectly rational Stoics, perhaps, but it would be odd to call them 'posthumans' at all since the term 'human' would lose its meaning. It is even doubtful if this extreme form of Stoicism would be *possible* for any entity that possess the capacity of valuing and that *engages* with the world. Again, transhumanists could achieve this kind of immunity and invulnerability if they would be prepared to give up these axiological and emotional ways of engaging with the world. Surely, if they wanted to avoid this consequence, they could propose more modest forms of enhancing and 'fine-tuning' our cognitive make-up, without compromising our capacity to care and value. However, this would imply that posthumans retain a large degree of their ethical-axiological vulnerability.

6.2.8 Conclusion

From this brief and general exploration of posthuman vulnerability, we can conclude that posthumans, if they would ever exist, will remain vulnerable in various ways. Although there are limits to our capacities to predict precise vulnerability

[6] There are various techniques for this. Probably the best technique is thinking about your own death; this technique continues to inspire people today. But the technique does not necessarily mean that you renounce all valuing. Combined with Romanticism, the Stoic technique means that you try to find value within yourself. As Steve Jobs said when addressing students at Stanford University: 'Remembering that I'll be dead soon is the most important tool I've ever encountered to help me make the big choices in life. Because almost everything—all external expectations, all pride, all fear of embarrassment or failure—these things just fall away in the face of death, leaving only what is truly important. Remembering that you are going to die is the best way I know to avoid the trap of thinking you have something to lose. You are already naked. There is no reason not to follow your heart' (Jobs 2005).

[7] For similar reasons, the modern (roughly Kantian) view that puts emphasis on good will, rationality, and autonomy cannot be considered a solution; instead, our craving for more autonomy is itself a source of vulnerability: it creates psychological frustration and inter-human violence.

transformations (of the kind I discussed in the previous section), we have good reasons to conclude that the project of erasing vulnerability altogether must necessarily fail. This renders the 'human enhancement' project of transhumanism problematic, at least in so far as it aims at invulnerability.

For this reason, we may want to turn to present, more urgent questions—although in the previous section I have also argued why it remains important to concern ourselves with ethical-anthropological questions raised by 'science fiction'. In the next chapter, I will respond to present concerns with new information and communication technologies. How are they changing the face of human vulnerability?

References

Anders, Günter. 1956. *Die Antiquiertheit des Menschen (Volume I): Über die Seele im Zeitalter der zweiten industriellen Revolution*. München: Verlag C.H. Beck, 1987.

Bostrom, Nick. 2002. Existential risks: analyzing human extinction scenarios and related hazards. *Journal of Evolution and Technology* 9(1). Available at http://www.jetpress.org/volume9/risks.html.

Briggle, Adam. 2008. Real friends: How the internet can foster friendship. *Ethics and Information Technology* 10(1): 71–79.

Coeckelbergh, Mark. 2007. Violent E-games, empathy, and cosmopolitanism. *Ethics and Information Technology* 9(3): 219–231.

Coeckelbergh, Mark. 2010. Imagining worlds: Responsible engineering under conditions of epistemic opacity. In *Philosophy & engineering: An emerging agenda*, ed. Ibo Van de Poel and David E. Goldberg, 175–188. Dordrecht: Springer.

Coeckelbergh, Mark. 2011. Vulnerable cyborgs: Learning to live with our dragons. *Journal of Evolution and Technology* 22(1): 1–9.

Floridi, Luciano. 2002. On the intrinsic value of information objects and the infosphere. *Ethics and Information Technology* 4(4): 287–304.

Floridi, Luciano. 2008. Artificial intelligence's new frontier: Artificial companions and the fourth revolution. *Metaphilosophy* 39(4/5): 651–655.

Habermas, Jürgen. 2001. *Die Zukunft der menschlichen Natur: Auf dem Weg zu einer Liberalen Eugenik*. Trans. H. Beister and W. Rehg. *The future of human nature*. London: Polity Press, 2003.

Hogue, Michael S. 2007. Theological ethics and technological culture: A biocultural approach. *Zygon* 42(1): 77–96.

Houellebecq, Michel. 2005. *La Possibilité d'une île*. Paris: Fayard.

Ihde, Don. 1990. *Technology and the lifeworld: From garden to earth*. Bloomington: Indiana University Press.

Jobs, Steve. 2005. Commencement address delivered at Stanford University, June 12, 2005. Stanford Report, June 14, 2005. Retrieved 7 Oct 2011 from http://news.stanford.edu/news/2005/june15/jobs-061505.html.

Johnson, Mark. 1993. *Moral imagination: Implications of cognitive science for ethics*. Chicago/London: The University of Chicago Press.

Jonas, Hans. 1979. *Das Prinzip Verantwortung: Versuch einer Ethik für die technologische Zivilisation*. Frankfurt am Main: Insel Verlag. Trans. *The imperative of responsibility: in search of an ethics for the technological age*. Chicago: University of Chicago Press, 1984.

Kurzweil, Ray. 2005. *The singularity is near: When humans transcend biology*. New York: Viking/Penguin.

Lakoff, George, and Mark Johnson. 1980. *Metaphors we live by*. Chicago: University of Chicago Press.

Lakoff, George, and Mark Johnson. 1999. *Philosophy in the flesh: The embodied mind and its challenge to Western thought*. New York: Basic Books.

Moravec, Hans. 1988. *Mind children: The future of robot and human intelligence*. Cambridge, MA: Harvard University Press.

Nussbaum, Martha. 1986. *The fragility of goodness: Luck and ethics in Greek tragedy and philosophy*. Cambridge: Cambridge University Press, 2001.

Sloterdijk, Peter. 1999. *Regeln für den Menschenpark*. Frankfurt am Main: Suhrkamp Verlag. Trans. P. Beers. *Regels voor het mensenpark: Kroniek van een debat*. Boom: Amsterdam, 2000.

Stahl, Bernd Carsten. 2002. Information technology, responsibility, and anthropology. In *Proceedings of the 35th Hawaii International Conference on System Sciences*, 7–10 Jan 2002, Big Island, HI.

Verbeek, Peter-Paul. 2005. *What things do: Philosophical reflections on technology, agency, and design*. University Park: Pennsylvania State University Press.

Chapter 7
Ethics of Vulnerability (iii): Vulnerability in the Information Age

7.1 Bodies, Others, and Reality in the Information Age: Is the Internet an Anti-vulnerability Tool?

If technology is an anti-vulnerability strategy, then what does this mean for the information and communication technologies that shape our present world? In what sense are they 'anti-vulnerability'? I already suggested in the previous chapters that they have not been particularly successful in diminishing our vulnerability. For instance, we are now threatened by computer viruses and (other forms of) cyber-crime. More generally, I have argued that technologies always *transform* our vulnerability and create new vulnerabilities. However, in order to gain a better understanding of these processes, we need to first explain why and how information and communication technologies (ICTs) can be interpreted as strategies to diminish or even eliminate human vulnerability.

7.1.1 Disembodiment, Disengagement, De-socialisation, Virtualisation, and Transcendence

Contemporary ICTs can be interpreted as anti-vulnerability tools for at least five related reasons: they appear to disembody us, disengage us, de-socialise us, virtualise us, and let us transcend earthly existence. Let me briefly describe each of these dimensions.

Disembodiment. The body is a major source of human vulnerability. But it appears that when we are using current ICTs, we take distance from our body. It seems that we can travel without our body in cyber space. Temporarily, at least, our body does not bother us.[1] This might be experienced as a liberation. Some

[1] Note that, if this is true, it also implies that we lack the *positive* possibilities related to embodiment.

M. Coeckelbergh, *Human Being @ Risk: Enhancement, Technology, and the Evaluation of Vulnerability Transformations*, Philosophy of Engineering and Technology 12,
DOI 10.1007/978-94-007-6025-7_7, © Springer Science+Business Media Dordrecht 2013

transhumanists propose that we dispose of it altogether and upload ourselves. This would mean that we are no longer humans. The fantasy is that we are pure minds, souls without bodies.

Disengagement. In the real world, we commit ourselves to particular projects and engage with particular things and people. This creates vulnerabilities, since if we engage, we also value, and we can lose what we value. Our projects can fail. In cyberspace, however, it seems that we are released from this troublesome human habit: we can freely roam, surf, explore, and enjoy without having to commit ourselves. We *like* something or we do not like something, but there are no obligations. This seems to be Dreyfus's view in his book on the Internet (2001), in which he draws on Kierkegaard (1846) to argue that the new medium does not promote putting ourselves at risk. We surf in an 'aesthetical' mode, but there is no ethical commitment. There are only opinions.

De-socialisation. If our relations with other people render us vulnerable and often produce suffering—perhaps up to the point that we can talk about 'hell'—then disconnecting ourselves from the real social world seems to amount to yet another kind of liberation. By entering cyberspace, we can free ourselves from the suffocating and oppressive communal web in which we had to live. We can still get in touch with people, of course, but with whom, when, and how long is entirely up to us (i.e. it is up to the individuals who communicate), and we need not commit ourselves to any relation if we do not want to. We can set up 'communities', if we *like*, but there is no need or pressure to strongly commit ourselves or to commit ourselves forever. It seems that the individualist-contractarian dream comes true: the social is a matter of (rational) decision and individual choice. It appears that we become significantly less vulnerable socially.

Virtualisation. Implied in the previous dimensions is the idea that the new media make us virtual. It seems that the new technologies create a second world, the 'virtual' world, which creates possibilities and capabilities we did not have in the first, 'real' world. In the real world, someone may be ugly, old, weak, lonely, cowardly, and vulnerable, but in the virtual world, that person can be a beautiful, strong, and invulnerable hero, the perfect fighter and the perfect lover, an attractive and smart superwoman or superman. In games and virtual worlds, people can create their own characters, bodies, and identities. They can also have tools that do not exist outside the virtual world (e.g. a powerful weapon, a magic device). Furthermore, whereas in the real world one's identity is firmly tied to one's appearance, it seems that the online professional and private sites and social networks allow us to choose our own face, our own mask. We can present ourselves to others in the way we like. Once again, this is often experienced as a liberation, including a liberation from the real vulnerabilities we have.

Transcendence. This attempt to escape human vulnerability and transcend earthly existence does not come as a surprise: it can be regarded as an integral part of, and continuation of, a strong current in the religious history of mankind—in particular those religious that emphasise transcendence. Consider the main monotheistic religions here, which separate heaven from earth, the city of God from the earthly city (see e.g. Augustine). Salvation is at least partly—if not entirely—about being saved from suffering and from death. The large monotheistic religions present themselves a ways to salvation, that is, as powerful anti-vulnerability strategies. First there is the Fall, then there is salvation. But what is the Fall? In Genesis 3, the Fall is related to suffering and the earthly as opposed to paradise and the divine: it is related to awareness of nakedness and to shame, to animality (the serpent), to women, to sorrow and conception, to labour and to eating ('In the sweat of thy face shalt thou eat bread, till thou return unto the ground'[2]), and to mortality: Adam and Eve lose their immortality; they can no longer live in paradise and are highly vulnerable. We can also think of Plato's philosophy (his dualist metaphysics and epistemology which distinguishes between a world of ignorance, the shadows in the cave, and the real world, the forms, the truth we see when we get out of the cave—but also his view of the soul and the afterlife), Aristotle's philosophy (the prime mover is external to the world), and the post-monotheistic modern philosophical tradition, such as Kant's philosophy, which distinguishes between a transcendental, 'noumenal' realm (the world of ideas) and a phenomenal realm (the sensory world). Moreover, one could draw a parallel with Gnosticism and Manichaeism. Gnostic thinking sees the material world as imperfect or even evil (since it was not created by God but by the demiurge) and seeks salvation of the soul from that world. Similarly, those who use contemporary ICTs as a way to escape earthly vulnerability seem to assume a dualistic world view, according to which there is a split between an imperfect or evil 'real' world and a better 'virtual' word: the Internet becomes a *salvation machine*, which enables people to transcend earthly existence and find salvation in the eternal cyberspace. To borrow a term from a very different (pagan) religious tradition: cyberspace promises to be an Elysium where we can indulge in whatever we enjoy and live a blessed heroic life. In the Elysian Fields of the virtual, we can be invulnerable, immortal, and eternally happy.[3]

7.1.2 *Being Online: Why ICT Fails as an Anti-vulnerability Project*

While the concepts outlined in the previous section help us to understand how and why the new technologies and media can be interpreted as anti-vulnerability tools and especially why they are *used* as anti-vulnerability tools (i.e. with the aim of decreasing vulnerability), I would like to offer an alternative interpretation, which rests on different ontological and epistemological assumptions and which shows that even if ICT can be interpreted as a tool to disembody, disengage, de-socialise, virtualise, and transcend the earth, it cannot succeed and necessarily fails as an anti-vulnerability project.

[2] Genesis 3:19, King James Version.

[3] Note, however, that in general nature, religions and polytheistic religions put more emphasis on immanent spirituality and that religions often contain both transcendent and immanent elements. For example, in Christianity God is transcendent, but becomes immanent in Christ, the one who *incarnates* God.

When We Are Online, We Don't Leave Our Body at Home. Even when we use information technology to roam in 'virtual' space, we remain embodied beings. Our mind is not disconnected from our body when we are online. The way we perceive, act, and think 'online' is still profoundly enabled, shaped, and limited by the embodied beings we are. Even if we create and use a 'virtual' body (an avatar), we can only use it as a virtual *body* on the precondition that we are embodied. Of course we may be less aware of our body, which can be problematic and even lead to (more) health problems. But the body-mind is still the experiential 'standpoint'. We do not become bodiless beings. Of course our environment changes, and in a sense we are 'extended' by the World Wide Web. But even if there is a sense in which the world becomes our body, this is still a form of (globalised) *embodiment*. Moreover, the globalisation of vulnerability does not decrease but rather increase vulnerability. A global 'body' implies much more vulnerability. But even if my 'skin' spans the globe, even if the skin of someone in another part of the world becomes also *my* skin, my experience of suffering is still a human-bodily experience of suffering. Finally, the very idea of 'uploading' one's mind presupposes a Cartesian dualistic view, according to which the mind is already separate from the body (the only thing uploading does is to disconnect what was already almost separated anyway: it involves cutting the cable so to speak). It assumes that we are robot 'bodies' (hardware) in which minds are downloaded or implemented. But we are body-minds and inseparably so. Therefore, uploading is not a possibility for beings like us, at least if we want to stay conscious and alive.

Condemned to (a Basic, Existential Form of) Engagement. There is little doubt that information technologies can lead to moral disengagement and the absence of commitment. However, they do not *necessarily* block engagement and commitment (for instance the press Kierkegaard criticised has also motivated people to engage themselves, and the same can happen with Internet), and at a more fundamental level, we are engaged and committed 'by nature'. Even if we do not explicitly commit ourselves to a particular project, as intrinsically moral beings, we are condemned to relate to others and to the world, and we are condemned to value—that is, we are condemned to existential vulnerability. We cannot but put ourselves at risk. Even if we try to live in a purely 'aesthetical' way, even if we try to 'practice' nihilism, we cannot *live* in this way. Therefore, rather than saying that 'the Internet' hinders 'engagement', it is important to study the precise ways in which specific programs and applications shape and frame our relation to the world and to others. But what we do 'online' (and offline) always matters to some extent, to others and to ourselves. In this sense engagement, or at least a basic form of it, is inevitable. We cannot escape risk by connecting to the World Wide Web. We are still 'in' the world, for being online is still a kind of 'being-in-the-world'. Moreover, the Internet is not an amoral sphere or an asocial sphere; we still have obligations to others. The Internet is not a sphere where 'anything goes'. This is because it is still a human, that is, moral-social sphere which is moral *before* we start reflecting on morality (see also Sect. 8.2.2).

The Social Is Everywhere. Those who look for a realm in which they can hide from the social, search in vain. 'Being-with' (to use a Heideggerian term) is our existential condition. In contrast to what contractarians assume, we are always already social. The social is not (entirely) up to choice. We might put on a different mask online (and this may be easier online), but we cannot escape the social game. And with whom we get in touch and how is not entirely up to us either. Like in 'real' life, sometimes we also 'meet' people we do not want to meet, do things we may not want to do (e.g. follow rules we do not want to follow), and receive an identity we might not fully want to have. We never fully control our identity, since this identity is always to some extent a 'third-person' identity. The 'they' (to use another Heideggerian term) is everywhere. The 'they' crosses the online/offline border. The social stretches across the real/virtual border, if there is such a border at all (see below). We always encounter the real—whatever technology or medium we use. The people we communicate with online are (usually) real people. Moreover, on the Internet, there is also oppression; social networks might even create a hyper-sociality in which we are constantly living under the gaze of others. This creates a heightened (awareness of) social vulnerability. Finally, the 'real' people we meet offline cannot be reached without some mediation and are not entirely 'authentic' either; we always wear a mask and we always perceive others through our cultural and personal glasses.

We Have Always Been Virtual. There is no ontological split between a virtual world and a real world. What we do 'on' Internet is as real as anything else we do, and what we do offline always had a virtual quality: the world in which we live is a perceived and constructed world—constructed by our scientific imagination, for instance. We also construct ourselves and are constructed by others. We do not have unmediated access to a 'real', noumenal, transcendent world. And technologies also play a role in offline or virtual life. Thus, the distinction between online and offline must be questioned. There is no separate cyberspace. There is one experiential space. The very idea of cyberspace is too dualist since it suggests such a separate space. It presupposes that you can travel to it and in it, that is, it presupposes that you leave home. But you never leave home in the first place; you always remain in the real world. You dwell online and offline at the same time. You are the opening in which the world 'worlds'. You are the actor, but the actor is at the same time virtual and real. This was always already the case, also without the Internet. You cannot leave the socio-ontological stage, you cannot leave the theatre. There are real violations and real vulnerabilities which you have to deal with. If you want to become a hero, you have to *act* heroically—online or offline.

Immanent Spirituality. If there is no 'other' world, no world 'behind' this one or no real cyber world, then there is no salvation, no immortality, and little room for transcendent spirituality. Then we must turn to this world (if there is another one at all—to use the phrase 'this world' seems to presuppose that there is another one). But this need not imply scientific materialism or nihilism. The objectification of the world (which is turned into 'the universe') and the related objectification of human vulnerability presupposes a deeply dualistic and in the end *nihilistic* world view. Both scientific materialism and nihilism presuppose that the world is secularised, that it is a void without meaning, without value, and without anything sacred. But we could reject this way of experiencing the world (but not *easily* so) and turn to immanent spirituality. We could follow the soldier of Borges's fable 'The Immortal' (1949), who stopped longing for immortality and eternity since it was too boring and instead decided to become mortal again, restoring his peace and joy. But recovering and affirming the earth and our blood, affirming the human drama—that is, affirming our human vulnerability and affirming the possibility of *meaning* and significance—should not be interpreted as a turn from the 'virtual' to the 'real', a turn from 'online' to 'offline', or a turn from 'technology' to 'authenticity'. 'Earth' is also the World Wide Web, and our 'blood' also runs through the computer networks on which we depend. If we must search for spirituality at all, it is not to be found outside the body, home, and world we live, and as it happens, this body, home, and world is dramatically and tragically changed by new information and communication technologies. But we are a drama and we are a tragedy, and when good and meaning emerge, they emerge within that existential space.

This proposal for a moral immanent spirituality can be read (i.e. reconstructed) as a sympathetic but critical response to Dreyfus. In his recent work, Dreyfus also seems to turn to a more immanent spirituality. In *All Things Shining,* Dreyfus and Kelly search for meaning in our 'secular' age and turn to a kind of polytheism: we are asked to respond to the sacred that is there and to the calling of the gods, to discover the 'shining' of things we intensely engage with (Dreyfus and Kelly 2011). I sympathise with Dreyfus's and Kelly's anti-nihilistic project (although I would not formulate it in theistic terms). However, when it comes to applying their suggestions to our contemporary condition, I believe Dreyfus is too pessimistic about technology, including current information technology. Largely due to his adoption of Heidegger's view of technology, he criticises our 'technological' way of thinking (and rightly so), but does not seem to differentiate between different technologies and different *uses* of technology. Although he does not explicitly say so, if we combine his view in *All Things Shining* with what he says in his book on the Internet (Dreyfus 2001), it seems that he would suggest that the Internet necessarily involves a 'technological' way of thinking that removes all meaning from the world. I agree that Internet often promotes the kind of autonomy and the kind of nonskilled type of activities Dreyfus rightly criticises. But why would the World Wide Web and cyberspace close off *any* possibility for skilled engagement and for the discovery of meaning? Perhaps there are 'shining things' to be discovered and meaningful, skilled engagement to be enjoyed in the practices and lives mediated by new information technologies, that is, in *our* lives today. Moreover, as I said although some things might allow for more (skilled) engagement than others, there is a kind of basic, existential engagement one cannot avoid. Dreyfus and Kelly write that we abandoned the gods (p. 222), but how far can we turn away from the sacred? How *transcendent* can the gods be? The narrative of turning away from the gods (or the gods turning away from us) presupposes that we (or the gods) can attain such a distance in the first place. But we have never been (completely) secularised. There is already light; things are already shining. As Dreyfus and Kelly rightly suggest, recovering the sacred is about discovery (p. 216), not about *giving* meaning.

My suggestion that we better see the (online and offline) world as impregnated with meaning through our social and embodied engagement with it does not imply the view that everything that happens in this world is good. My interpretation of being 'online' as being-in-the-world, as being-with (Heidegger's *Mitsein*), and as part of a non-dualistic whole without transcendence does not exclude the evaluation of particular ways of perceiving and acting. In other words, it does not exclude the possibility of ethics. However, by 'ethics' I do not only mean rules for conduct. Of course, within a particular game and a particular theatre play, there are some normative constraints and rules (if the social life or a practice is a game that does not mean that anything goes, the game has its rules). If thinking about these rules is ethics at all, then this would be an ethics of 'right'. But by 'ethics' I mean an ethics of 'good', in the following sense: we can still make distinctions between ways of life and practices that significantly contribute to, and embody, more engaged, more social, and more spiritually wholesome ways of relating to the world and to others, and other ways of life that do this to a much lesser extent. At a fundamental existential level, we are all related. However, there are particular ways of doing and thinking (habits) that make us experience ourselves and our relation to the world in such a way that we behave and think in a way that denies our fundamental existential relationality. From this perspective, some things Dreyfus says about the Internet (Dreyfus 2001) are relevant: perhaps some ways of doing (online or not) *are* not good, but not because they 'take place' 'on the Internet' (as if that were a *separate* place) but because they deny our relationality by discouraging commitment and engagement. Although 'the social' cannot be damaged, particular forms of the social *can* be damaged and *are* vulnerable, such as the rather vulnerable public sphere Dreyfus and Kierkegaard want to have. We cannot destroy social space as such (just as we cannot destroy physical space as such) and we cannot 'undo' the basic existential sociality of humanity, but we can destroy particular forms of social space, particularly architectures of social space and their transformations. We can also contribute to the emergence of new forms of social space—even if we cannot completely 'design' those spaces. We can try to steer the transformations in different directions.

For example, violent computer games are often controversial: some oppose them and think they are very dangerous. But how can this opposition be justified? What exactly is wrong with them? If we assume that the distinction between real and virtual is (1) valid and (2) morally significant, then we arrive at one-sided views, such as the view that virtual violence is unproblematic because it is merely virtual or the view that virtual violence has consequences 'in reality'. Neither view is very helpful for thinking about the problem. Instead, the alternative view I defend does not start from dualistic premises, and it evaluates the moral quality of the action—in particular the *habit* and the kind of skilled engagement involved. Consider what happens in violent games: gamers often shoot over and over. In a sense they are *trained* to shooting. My proposal is to disregard the question if this shooting is online or offline, but evaluate the moral quality of the habit and skill of shooting. One could try to justify why that *habit* is bad—by whatever means it is created.

Let me try out a very brief phenomenology of shooting in order to illustrate my approach. The gun allows me to fight with someone at a distance. In contrast to martial arts, there is no need for bodily contact. I can kill without touch. This means that I do not need the skills of the (traditional) martial arts; the main skill I need is aiming and perhaps some basic maintenance skills. (Note that in computer games, even the skill of aiming can be automated: so-called auto-aim or smart aim means that the computer assists you with aiming. And in contemporary warfare, the soldier can be enhanced by various electronic devices, including auto-aim. The technology is expected to move from video games to the battle field.)

Moreover, killing without touch also means that I can kill at greater distance. In this sense, the gun 'creates' distance between me and the other. It does not only create deskilling; it is also a *disengagement* and *de-socialisation* instrument par excellence. It denies the relation between me and the other and it denies the other's sociality. The other becomes a *target*. The gun does not differentiate between humans and things; both are merged into one single category: the target. This procedure of dedifferentiation and dehumanisation allows me to shoot at the person, who no longer appears as a person but as a target, an object. The gun is an objectification technology. It makes the violence possible.

We can use Levinas to support this point. Levinas argued that an ethical demand arises from 'the face of the other': the face of the vulnerable other says, 'Do not kill me'. The 'nakedness of the face' (Levinas 1961, p. 74) renders killing impossible. I cannot kill if I experience in a face 'the ethical resistance that paralyses my powers and from the depths of defenceless eyes rises firm and absolute in its nudity and destitution' (pp. 199–200). In this sense, shooting-thinking makes *the face of the other disappear*.[4]

Thus, a crucial condition of possibility for gun-mediated violence both in computer games and in the 'real world' is not the information technology or the game, but the technology-skill and the technology-habit of shooting. Shooting a gun is a way of perceiving and thinking as much as it is a way of doing. And if our thinking becomes a shooting-thinking, then violence becomes possible. Shooting-thinking makes the other disappear. In the eye of the shooter, the 'other' does no longer appear as a human, as a social other, and as someone-who-is-vulnerable. In the perception of the shooter, the person does not even have the vulnerability of an object, since once the person becomes 'the target'; it is 'something that has to be hit', not 'someone who can be wounded and who can die', not someone with a face. This switch allows the shooter to kill—allows *everyone* to kill. Once others start appearing as targets, as objects that stand in my way and need to be hit, removed, and terminated, then I *become* the gun and the bullet. When I aim at what has become the target, the bullet has already been released, so to speak. This is why training the habit of shooting at others is bad, online or offline: it denies the humanity of the one who may get shot and of the one who shoots. It denies our mutual

[4] Sometimes people who want to act violently towards others (e.g. towards a prisoner or someone who will be executed) cover the face of the person before they act. They first 'remove' the face; the victim then appears as an object rather than a human person.

interdependence and vulnerability. It denies the shared blood and the shared wound. If you bleed, I bleed.

7.2 Beyond Cyber Security

My view of the relation between vulnerability and technology has consequences for thinking about security, for example, 'cyber security'.

Cyber security is a big challenge today. We try to protect our programs, computers, and networks from access and damage. We try to prevent hacking, the penetration of our computer system without our permission. We try to prevent our computer from being taken over. It might become part of a botnet and then used for malicious purposes. Typically for cyber security risks, vulnerability change is fast: there are always new threats and new risks. We take counter measures, we create firewalls and antivirus software to keep out the viruses, the spyware, the 'worms', and the 'Trojan horses'. But software development is fast; often countermeasures cannot keep up with the changes. It seems difficult to protect our computers and other electronic devices from cybercrimes. Our experience cannot keep up with the changes: we are short of words. Furthermore, at stake is not only the security of individuals but also of companies and statues. There is industrial espionage and states also engage in cyber spying. States may even try to cause damage: they may use hacking as a means of 'cyber warfare'. Military organisations start to recognise cyberspace as a new domain yet one that is similar to land, sea, and air: it is a new (potential) 'battlefield'.

Related to this issue of security is the topic of *privacy* (and, related, the issue of individual autonomy). If we do not want others to consult or steal our data, it may be because we care about keeping some data private. We see this as part of our personal integrity. We define some data as 'personal information' which we like to control and protect. Sometimes we willingly give our data to others. But we like to decide when and to whom we give personal information. We do not want others to access or use our data without our consent. However, states and especially companies often collect, store, and use data about us which we regard as 'personal information', without always asking our permission. Thus, we value privacy and individual autonomy, but these values are not always respected—they seem to be threatened by the new technologies and by those who use them to invade our private domain. (A different way of putting the threat to autonomy in the case of data collection: we become a kind of new slaves—we become data workers or data factories, producing data for large multinational companies. We become raw materials, we are *mined* for data.)

In sum, new information and communication technologies seem to create new kinds of risk and vulnerabilities, which are usually interpreted as a problem of *security* and privacy, often also as a problem of autonomy, and with which we try to cope by developing particular countermeasures to *protect* us and our technological systems. Let me further analyse the way we understand our new vulnerabilities and explore an alternative way of framing the problem.

7.2.1 Dualist Views of Cyber Security: Autonomous Castles and Their Firewalls

The standard way how we perceive, interpret, and respond to security (and privacy) threats related to ICTs can be articulated by using the metaphor of the castle, a well-known defence structure originating in early medieval times, when certain people used it in response to their (offline) security problem. Today we understand and build our information systems as 'private fortified residences'[5] which we defend against external attacks by malicious agents. We turn our computers and other systems into e-castles. We build firewalls and take other measures to keep the enemy out. In this way, we hope, and we render ourselves less vulnerable. We gain security and privacy within the walls of our castle, the fortified residence of the autonomous individual and the authentic self.

This way of responding to the problem assumes the following beliefs. First, it assumes that we take a fixed, rather passive position, waiting to be assaulted. We prepare for the raid by taking countermeasures, but we do not move beyond the walls of our castle. The enemy moves; we are immobile. We only act within the walls of our (computer) system, not in (the rest of) cyberspace or in the larger online/offline world. Second, we assume that there is a strict separation between 'inside' (the castle, the safe zone, the private) and 'outside' (the not castle, not safe, not private). Security (and privacy protection) is aimed at maintaining this separation. To let the darkness in would be the horror. It's me-and-my-computer against the (bad) world. Third, it is assumed that good and bad, light and darkness, can be separated easily: 'we' are good and 'they' are bad (the malicious software, the malicious hacker). We assume that there are clear moral distinctions. We protect the innocent virgin (our data, our computer, our private life) who should not fall in the hands of 'the bad guys'. Fourth, even if there might be some good people out there, we assume that if the enemy (e.g. the computer virus) arrives at our gates, we can recognise him—that is, we can recognise him *as* an enemy, we can recognise that he is an *enemy* in the first place. For example, we assume that our security system can distinguish between viruses and other (pieces of) software, 'good' code and 'bad' code. Fifth, therefore we assume that we can distinguish between 'a security situation' and 'a non-security situation' and between an attack and a non-attack (say a friendly visit). And last but not least: we assume that by retreating into our castle we *are* protected. Behind the firewall we can live in peace, as long as we take sufficient protection measures (or pay others to take care of that for us). To take up again the pseudo-medieval (Romantic) imagery: on his throne sits the noble autonomous individual, well protected and—this is perhaps an additional assumption—all *powerful* when it comes to deciding who is welcome. We can decide who we let into our castle, our private grounds.

---

[5] I use the Wikipedia definition here, available at http://en.wikipedia.org/wiki/Castle

Whether or not the world could *ever* correspond to this image of security (or indeed *could* ever correspond to it), these assumptions are rather problematic in the information age. First, as users of information and communication technology, we are neither immobile nor confined to a specific space. Rather, the World Wide Web allows us (and our data) to move 'everywhere'. This is not something that merely happens to us but has to do with our activities. If we are at risk, it is because we *put ourselves at risk* by our activities mediated by the World Wide Web. In line with historical usage of the term, we must conclude that risk is unavoidable when we explore the electronic seas and lands of cyberspace. Once we became 'Internet explorers' (indeed there is a sense in which we *become* the software we use; compare with becoming the gun), we also incurred the risk associated with these voyages. There will always be the possibility of electronic shipwrecking. We can *insure* ourselves against this, perhaps, but not (entirely) *secure* ourselves against this vulnerability.

Second, there is no privacy and no separate bunker where we can hide. Once our computer is connected to the Internet and the web, it also becomes a very *part* of that network and that web. In a sense our computer or phone is always *already* a bot in a botnet; without the computers and servers that make it up, there is no Internet. Maybe the network becomes ontologically prior to its nodes. Maybe at some stage, it will become less meaningful to speak of computers as constituting distinct objects or as occupying distinct spaces. Maybe it will become less meaningful to speak of 'computers' at all when it comes to describing information and communication technology. Don DeLillo writes in *Cosmopolis*: 'Computers will die. They're dying in their present form. They're just about dead as distinct units.' Moreover, if we depend so much on the Internet for our activities, we cannot hide in 'offline' world. Increasingly, our world becomes at the same time online and offline. Floridi has coined the term 'onlife' to refer to the blurring of the distinction between online and offline:

> Information and communication technologies (ICTs) are building a new habitat (infosphere) in which future generations, living in advanced information societies, will spend an increasing amount of time. In the infosphere, it is progressively more difficult to understand what life was like in pre-digital times and, in the near future, the very distinction between online and offline will become blurred and then disappear. The phenomenon is variously known as "Ubiquitous Computing", "Ambient Intelligence", "The Internet of Things" or "Web-augmented things". GPs [global positioning devices used for navigation] are a good example of this convergence: asking whether you are online when driving a car while following some GP's instructions updated in real-time is becoming progressively less meaningful. We already live mostly onlife. (Floridi 2011, p. 477)

Third, given our involvement in electronic activities, the distinction between good and bad is blurred too. This is partly so since, as I said, we take risks. This makes us co-responsible for what happens. Moreover, when is someone a 'hacker', and when should we call his actions—often it still is a 'his'—'bad' or 'morally wrong'? Few Internet users are intentionally malicious. But there is a grey zone. The same hackers may do 'good' and 'bad' things, sometimes in order to reveal vulnerabilities (so-called 'grey hat' hackers). In the latter case, they seem to perform a function similar to that of vaccination: their intervention is meant to make the system stronger. By showing the weak spots, the action is meant to help the managers of the system decrease the *vulnerability* of the system and thus our vulnerability. And most of us use software from large ICT companies because we gain advantage from using their software (e.g. social networking software), even if we know that they use our data to *their* advantage (e.g. sell our data to other companies who use it for advertisement and marketing purposes). In so far as we contribute to this organised system of data prostitution, the difference between the good and the bad is blurred. The darkness is not only 'out there', it is already in our activities.

Fourth, we and our systems cannot always recognise viruses or other malicious software. 'Bad' code can be disguised as 'good' code. The enemy may be dressed up as our friend. By the time we can identify one disguise *as* a disguise, new disguises are already in use. There are endless processes of unmasking and masking.

Fifth, this also means that there is no 'peacetime'; we are always at war. Perhaps there are only ceasefires. We are living in a continuous state of alarm, a 'risk society' (Beck 1992). Both our hardware and our bodies are under stress.

Finally, this also implies that we are never safe and that we are not all-powerful. We do not fully control our security. Information life does not work in terms of consent. Of course individualist-contractarians would like it that way (e.g. we still use instruments such as consent forms and agreements), but in practice we already put ourselves in the hands of machines, of software, of others, and of companies. Often we have 'no choice' but to agree, especially if want to use software that is compatible with software other people use (the lazy way of reaching compatibility is to use the *same* software). We do not have full control over our data; we are already giving it all to the cloud. Our 'onlife' is private and public at the same time.

Note also that one reason why we—as users—are not very powerful in the first place is that the information technologies we use and the worlds they create have become so complex and opaque that most of us have delegated the management and work on the technologies to others, to experts. But this also means that we delegate *power* to them. Moreover, this also means that *they* have and develop the skills and know-how, whereas *we* lack the skill. As Dreyfus and Kelly rightly complain, in the 'technological age', we are treated as children. They write that 'advances in technology have diminished the importance of specialized skills in contemporary life. (…) "Even a child can do it!" is the mantra of the technological age' (Dreyfus and Kelly 2011, p. 212). Thus, by wanting 'easy' devices that hide all their 'inner' workings to us, we have gained easy use, but we pay a high price: the loss of power, the loss of autonomy, and eventually the loss of engagement and meaning. Paradoxically, by trying to cover the world with our web, it got out of our hands. Now we are waiting for the next feeding and the next toy. Our security and vulnerability condition becomes that of a baby. We think we are living in a castle, but we are in the cradle, crying for mother Microsoft, father Facebook, aunty Apple, sister Twitter, and (big) brother Google, who *know everything* and who will *take care of us*.

Given these problems with the assumptions we make about security, and indeed with the very practices we are concerned about, we need new concepts and, more generally, new ways of thinking about risk and vulnerability in 'cyberspace' (which does not exist as a separate space in the first place—see also the previous section).

7.2.2 A Non-dualist View of Informational Vulnerability: Ubiquitous Vulnerability in Cyber Ecologies

A non-dualist view of 'informational' vulnerability could use the language of extension to make clear that we have 'spread out' into cyberspace and therefore have *rendered* ourselves vulnerable. If we have extended ourselves by means of the new information and communication technologies, then we have also extended our vulnerability: dispersed into the World Wide Web, we are now 'globally' vulnerable. Our nervous system has been extended to the Internet (something McLuhan had foreseen), our skin has been stretched and spread out over the electronic globe, our bloodstream has been extended into the electronic streams that cover the planet.

We could also take up the metaphor of travel and use the concept of 'nomadic' security: we do no longer have a fixed base, we move all over the place, and hence what we have to defend (e.g. our integrity or our identity) can no longer be captured by the metaphor of a castle but with that of a car or a caravan. People and bites both move around the globe, and (good and bad) things can happen to the caravans of people and information. Vulnerability then pertains to the vulnerability of the caravan, not to the vulnerability of an immobile fortress. This concept does more justice to the idea that we render ourselves vulnerable, that we put ourselves at risk, by 'going places'.

However, the language of extension is still too dualist: it presupposes that there is a 'something' that has been extended and that there was a *separate* cyberspace or infosphere in the first place. The concept of nomadic security faces a similar problem: nomads still make a strong distinction between on the one hand the 'their' people, things, bits, etc. and other the other hand the (people, goods, and bits of the) outside world. They travel, but their caravan is their mobile fortress, their mobile identity, and their collective authentic self. Our cars become mobile fortresses and strongholds of personal identity. Some people want to have their car bulletproof: it's again 'me' against 'the world'.

Less dualistic terms, therefore, to replace the concept of 'extended vulnerability' or 'nomadic security' are '*ubiquitous* vulnerability' or '*ambient* vulnerability'. If we live in a cyber ecology (which is neither merely 'electronic' nor merely 'biological', 'physical', or 'mental'), then vulnerability is ubiquitous too. If the world has become our body-mind, then everything is dependent on everything else, and we have to think of risk and vulnerability in *holistic* terms. A 'local' computer problem can have worldwide consequences. What happens in the parts constitutes what happens in the whole system and vice versa. Then there is no longer a sharp separation between 'me' and 'the outside world' (a gap which then may be bridged by means of 'extension'); rather, there is one world of which I am part. I have a world and the world has me.

Perhaps this means that we must replace the central concept of 'security' as well, since it seems to presuppose the dualistic thinking that hinders us from fully understanding vulnerability in the information age. Maybe 'health' would be a better term, provided that health is understood ecologically and holistically. The integrity of the parts depends on the integrity of the whole. In the information age, 'my' vulnerability as an 'individual' 'computer user' (both terms make a lot less sense seen from this perspective) entirely depends on the vulnerability of the systems in which I partake. We share information and therefore we share vulnerability. To connect is to put yourself at risk as part of a larger whole. The concept of 'security' measures is replaced by the concept of making the whole healthier, involving the development of a good *immune system* that is not 'individual' but shared by the whole information-technological ecology and also the development of a better 'lifestyle' (remember that we put *ourselves* at risk, we do risky things). Some practices may be harmful to the ecology as a whole and should be at least discouraged. Moreover, we have to reconsider if we want to delegate the management of our health to the ICT doctors of this world and their helpdesks and to the 'pharmaceutical' computer security industry or if we want to develop systems or support existing tendencies of self-healing and communal healing. We could also think about how we can provide more opportunities for meaningful and skilled engagement with the technologies and with the world, that is, develop transparent systems that treat us as adults rather than babies or patients. This may not dramatically reduce vulnerability, but it may give us *a kind of* vulnerability, that is, a kind of *existence*, that is more worthy and dignified.

Note that this holistic, ecological approach should not be understood as implying the thesis that we now face an entirely *new* situation or phenomenon, produced by the new technologies. As humans we were always already part of a larger whole, and we were always already dependent on that larger whole. Our vulnerability was always ecological and holistic from the start. What differs in the information age is the precise character of the whole: its nature is now not merely biological, physical, mental, and so on but also electronic, informational, etc. And when the whole changes, the parts also change. Humans change. Our security and health can no longer be understood only in terms of our biopsychological condition; our condition and our integrity are now also dependent on the informational systems of which we have become a part. There is no such thing as 'informational security' if that term refers to an entirely separate kind of vulnerability and to entirely separate kinds of risks. Instead, existentially and metaphysically, we are at risk as physical-biological-psychological-informational beings. A threat to an informational system is also a threat to *us*. The possibility that our software, our computers, our phones, and our navigation systems may be violated is the possibility that *we* may be violated.

Let me explore the phenomenology of these kinds of violations and the vulnerability involved by using the image of gravity and antigravity struggle. When it comes to existential vulnerability, a crash of our computer is comparable to a fall of our body. In both cases, we try to keep upright, we wish to stand out in our verticality, and we actively seek verticality (i.e. we take measures in order to reduce our vulnerability), but there is always gravity. We still remain vulnerable and mortal, in spite of our computers and our networks. Moreover, because we move around in new ways we incur new risks, we become vulnerable in new ways. We have to fight computer viruses and our systems can crash. We become dependent on worldwide networks. When we grew up, we met new kinds of resistance. A Dutch proverb says: high trees catch a lot of wind. We are creating and becoming enormous trees, which are spreading out their branches everywhere. This creates new vulnerabilities. The infosphere has its own winds. There is resistance. We run, but sometimes the wind blows hard in our face. Existential gravity and resistance means that in spite of our anti-vulnerability strategies, we remain at risk, and we remain vulnerable, often in new, unforeseen ways. And sometimes we can no longer stand up and start up.

For some people, the wind blows harder than for other people. How vulnerable you are, how vulnerable you become, and what you can effectively do to reduce vulnerability depend (among other things) on how powerful you are. This is the topic of the next chapter. But let me first end this chapter by briefly discussing the relation between vulnerability and new technologies in medicine and health care. What if we cannot get up?

7.3 New Vulnerabilities in Medicine and Health Care

In order to further clarify and complete my view of the relation between vulnerability and (information) technology, let me discuss a more concrete field of application: medicine and health care. This also serves to make clear what is at stake in ethics of technology: it is about games and software, for sure; but it is also about life and death.

7.3.1 Wired Patients: Cyborgs in the Intensive Care Ward and Robots for the Elderly

When we think of the information age, we tend to have in mind healthy, young, and highly active people who have all human capabilities and who use computers, smart mobile phones, and other electronic devices in their leisure time and at work. But information and communication technologies are also increasingly used in medicine and health care. This means that these technologies also shape our experience of illness, old age, disability, and dying—indeed experiences at times when we typically become (more) aware of our vulnerability.

Technology plays an important role in contemporary medicine and health care. If we are seriously ill and taken to hospital, we soon become *wired* patients, often literally. If you are looking for cyborgs, you do not need to watch science-fiction movies; just visit a modern hospital. You see them in the intensive care ward and in the operation room. People and machine are tightly connected, and people depend on the machines for their health. But information technology is also a part of non-intensive care and medicine. If we enter the consulting room of our local medical doctor (general practitioner), we do not only meet a human being but we also enter a world of artefacts. Among these artefacts, the computer takes a central place in the practice: the computer and the doctor collaborate, working together to construct the diagnosis and to propose a treatment. Again we meet a cyborg. Moreover, information and communication technology is increasingly used in nursing, both in care institutions and in home care and telecare (remote care). Electronic devices help to organise the care and monitor the patients and the elderly people. Intelligent exoskeletons are used for rehabilitation. Robots are used to communicate with patients at a distance and to monitor elderly people who live at home.

One way to interpret the widespread use of technology in medicine and health care is to see it as a specific way of coping with vulnerability and of trying to reduce vulnerability. The technology is meant to reduce the vulnerability of the patient or elderly person. For a patient in a critical medical condition, being connected means having a higher chance that action is taken when something goes wrong. For a general practitioner, being connected means having more and fast information available when a diagnosis and treatment advice has to be made, thus reducing the chance that the condition of the patient gets worse (and, at best, increasing the chance that it gets better). For a paralysed person, information technology may make the difference between being able to communicate and not being able to communicate or between being able to move a limb and not being able to move it. For an elderly person, electronic devices and robots can make her less dependent on human care and more autonomous in that sense, or it may make her feel less worried that when something goes wrong, no one and nothing notices.

Given these obvious benefits, we might conclude that ICTs in medicine and health care reduce vulnerability and are therefore to be recommended, to be welcomed, and to be used without hesitation. It seems that when we find ourselves in a vulnerable medical condition, we must hope that those who treat us and care for us use all information-technological means possible to cope with the wounds, the limitations, and the pains that have fallen upon us and (often literally) have pushed us down to the ground, down to the earth to which we do not wish to return (yet). It appears that when we feel the gravity of human vulnerability in illness and/or old age, we have good reasons to shout: 'Connect me!'—afraid as we are that our lines and dots might fade away from the screen, that we might go offline forever. Who does not want to walk again? Who does not want to be *resurrected*?

However, while this hope is understandable and (in many cases at least) not entirely without foundation, it is also important to realise that the new technologies (1) never completely remove any limitations or all vulnerability (including mortality), even if they restore us to 'health' and (2) do not only reduce particular vulnerabilities—and not only treat us but also *enhance* us in this sense—but also create new dependencies and (related) vulnerabilities. Let me explain and support these claims in the next section.

7.3.2 New Vulnerabilities: Objectified by Technology, Disabled by Nature

The first claim is clear enough: even if we use these technologies, there is no 'zero vulnerability' and we remain mortal. (With regard to the technological future, a few transhumanists disagree with the latter part of the claim, arguing that technology will make us immortal in the future, but I leave aside that position. Note also that many religions believe in an afterlife, but this is always a life *after death*—after 'biological' death, the death of the body.) Thus, information technology may reduce vulnerability, but it cannot make us invulnerable. At best it can make us healthy, but health is a temporary state; there is always the possibility of illness and death.

The second claim, however, needs more explanation. In this chapter and in the previous chapter, I have already made a general argument about information technology and why it fails as an anti-vulnerability strategy—it creates new vulnerabilities now and in the future—and here I would like to explore what this means for medicine and health care in particular: in what sense do ICTs in medicine and health care make us more vulnerable or vulnerable in a new way? And in what sense do they make us *less* healthy?

Dependence. To the extent that our health becomes dependent on ICTs, our vulnerability is directly dependent on the functioning of the software and hardware involved. If they fail, we might 'fail'. If they have a bug, we might get a 'bug'. This process can be described in terms of extension—if we extend ourselves, we also extend our 'vulnerability skin', our 'bloodstream', and so on—but it can also be seen in a relational-ecological way: our vulnerability depends on the vulnerability of whatever and whoever we are related to. If our health depends on information technology (and on those who operate and use the technology), then our vulnerability also depends on that technology and its users. In this sense, the vulnerability of a patient in intensive care is a kind of *cyborg vulnerability*. When we read 'cyborg', we think of science-fiction images. But this is only an exaggeration of what already happened to many 'wired' patients; it is hard to think of contemporary medicine and health care without information technology and other tools. We are ICT addicts.

Objectification of the Body. We are not only what we eat (another important relation) and with whom we eat; there is a sense in which we are also the *technologies* we use. We are not 'mere users'. To the extent that we depend on the technology, the modern health-care system turns health-care professionals and patients into parts of the system. The modern hospital becomes a health factory or a computer. Hospital processes resemble assembly lines of manufacturing processes (health is produced) and information processes (the sick patient is the input, the healthy patient is the output, and in between the patient is *processed*). Doctors and nurses resemble the robots of the assembly line or pieces of software (e.g. plug-ins) that work together to process the patient. In the information process, the patient becomes information or at least a 'container' and producer of information. We are treated as electronic devices. When we enter we are tagged. In order for us to be diagnosed, we are first connected. Data about our health conditions are retrieved, downloaded from us, we are *read*. We are monitored; we have to stay online. We are upgraded. New code is added; bugs are fixed. The result is that our vulnerability is not treated as human vulnerability but as the vulnerability of a thing. In particular, it is reduced to the vulnerability of our body, understood as a biological and/or informational system. Due to Cartesian modern thinking, modern medicine views humans in dualistic terms. Once we enter the hospital as 'patient', our body is presented to us and to others as an object, a thing. It is our machine, our robot. We come to think of ourselves as robots. In our experience of what is going on, we become alienated from our body. The body is something I (happen to) *have*. I bring it to the hospital to be repaired in the same way as I bring my car to the garage for a repair—or indeed my computer to the computer repair shop. I discipline my own body, for example, by means of sport or diet, in order to keep it *working*. The body is something that works for me, my robot-slave.

The result is that I come to understand my vulnerability in the following (dualistic) way: my mind is invulnerable and strong, but unfortunately I happen to have a weak, vulnerable body, which needs to be disciplined and managed. This modern understanding of the human being is not only echoed in modern ethics (e.g. Kantian ethics perhaps) but is also a condition of possibility for the transhumanist idea of uploading: only if I already understand myself as consisting of two parts, I can separate these two parts. I can leave my body and choose another one—or dispense with a body altogether.

However, the modern way of thinking must be seen as part of the problem rather than the solution: if we dissociate 'ourselves' from our bodies, this makes management of the body but also its abuse *and* disregard possible—both of which are unhealthy and constitute new vulnerabilities. If I see my body as something alien, something that is not 'me', then I may not care for it (in the right way). I may allow input and output exchanges, but use it as a processor or as a natural resource, until it is depleted and exhausted.

Isolation. ICTs make it possible that elderly people stay (longer) at home and do not have to spend many years in modern health-care institutions. However, this creates a particular kind of social vulnerability: elderly people may fear that when they get a robot and all kinds of intelligent devices, they see fewer human caregivers (professional or other). We can imagine a dystopia in which elderly people have to spend their days in 'care capsules' that are well connected to the Internet but where human contact and human *touch* is rare. This creates not only cyborg-type technological vulnerabilities because of the total dependence on the high-tech environment but also additional social and psychological vulnerabilities. Even if this worst-case scenario does not happen, both modern care institutions and home care already create and shape these vulnerabilities. People in such institutions have less social, emotional contact and intimacy with the people that are (were?) close to them. Elderly people fear having to leave their home, which means: leaving the building but also leaving their (social, emotional) world. With our new care technologies, we can try to enable people to stay in that world. However, to the extent that this is not possible (for certain people) or if it involves less care by humans (care robots, the care capsule idea), we will reduce bodily vulnerability but increase socio-psychological vulnerability, and if we hold a holistic view of health, we can even argue that these technologies might not even reduce 'bodily' vulnerability at all in this case since both types of vulnerability cannot be separated.

Financial and Political Vulnerability. The future of medicine and health care is highly dependent on the financial resources (that are made) available. In this sense, patients and elderly people are not only dependent on the high-tech health monitors and screens that surround them but also on the *financial* monitors and screens that co-write and co-produce their health future. Financial crises have indirect consequences for people's vulnerabilities, including health vulnerabilities. Patients also depend on political decisions—about finances and about other matters related to health care. In so far as ICTs contribute to the financial and political arrangements and processes we have to live with, they contribute to the creation and transformation of health-related and other vulnerabilities. Modern technology cannot be disconnected from modern society. In the next chapter, I will discuss the politics of vulnerability.

To conclude, new information and communication technologies are not neutral instruments for medicine and health care, but transform our vulnerability in various ways. Paradoxically, they at the same time reduce and increase our vulnerability.

This conclusion is not meant to be antitechnology in principle. In a sense, we are disabled by nature; we are incomplete and limited beings. We are neonates who are quite helpless without others and without technology. (At the same time, this is our strength compared to other animals: we excel in learning and adaptation.) Moreover, before these kinds of modern technologies, we were also dependent on technologies—*other* technologies. As relational beings, we were always ecologically vulnerable, and technologies have always been part of the ecology. What differs today—and what matters to ethics—is the specific current form of that ecological vulnerability, which depends on what and whom exactly we are related to and dependent upon today, on the precise shape of the environment we have created and are creating by technological and other means. A normative anthropology of vulnerability tries to understand and evaluate these ecologies, transformations, and architectures of human vulnerability.

In the next chapters, I will explore the political and ethical-aesthetical aspects of this project.

References

Beck, Ulrich. 1992. *Risk society: Towards a new modernity.* Trans. M. Ritter. London: Sage.

Borges, Jorge Luis. 1949. The immortal. In *The Aleph and other stories.* Trans. A. Hurley. New York: Penguin, 2004.

Dreyfus, Hubert L. 2001. *On the internet.* London: Routledge.

Dreyfus, Hubert, and Sean Dorance Kelly. 2011. *All things shining. Reading the Western classics to find meaning in a secular age.* New York: Free Press.

Floridi, Luciano. 2011. The construction of personal identities online. *Minds and Machines* 21(4): 477–479.

Kierkegaard, Søren. 1846. *The present age.* Trans. A. Dru. New York: Harper & Row, 1962.

Levinas, Emmanuel. 1961. *Totality and infinity.* Trans. A. Lingis. Dordrecht: Kluwer Academic Publishers, 1961.

Chapter 8
Politics of Vulnerability: Freedom, Justice, and the Public/Private Distinction

8.1 The Idea of a Politics of Vulnerability

There are several ways to understand the idea of a 'politics of vulnerability', each of which raises many questions at various levels of analysis.

First, the term can refer to political action that directly responds to the (potential) violation of humans, such as a politics of human rights or a politics based on the capability approach. Decisions and actions that promote human rights or human capabilities can be interpreted as contributions to a 'politics of vulnerability' that intends to decrease the vulnerability of those who are threatened by violence, poverty, and so on, by offering specific protections and by strengthening their capabilities.

Second, we can also conceive of a definition of politics as an anti-vulnerability strategy. Responses to violations of human rights and problems concerning capabilities are only a part of a broader societal strategy to minimise human vulnerability. This notion of politics as an anti-vulnerability strategy does not imply that state action is the only or best means to do that; it leaves room for different ideas on where the centre of political power should be and who (and which social institutions) should act to minimise that vulnerability. Moreover, it also leaves room for asking the question about the boundaries of the political. For instance, is politics an exclusively human affair? If vulnerability is not only about humans but also concerns animals or even things (technology, artefacts, natural phenomena, etc.), we might want to alter our thinking about politics and our political structures. What are the boundaries of the political community? What are the boundaries of the collective? For example, combining the approach of shared vulnerability with relational approaches to moral and political status (e.g. Coeckelbergh 2012; Donaldson and Kymlicka 2011), one could consider including (some) animals into the political community. And with Latour (2004), we could think about how to represent various entities, things, and events at the political level, by virtue of the role these entities play in creating and transforming vulnerability. Finally, whether or not this is a fruitful direction, existential vulnerability and risk need to be discussed as a public matter, since they *are* of public concern. But this urges us to reflect on what kind of

M. Coeckelbergh, *Human Being @ Risk: Enhancement, Technology, and the Evaluation of Vulnerability Transformations*, Philosophy of Engineering and Technology 12, DOI 10.1007/978-94-007-6025-7_8, © Springer Science+Business Media Dordrecht 2013 politics and what kind of *conception* of politics we need in response to this problem. We need to rethink the relation between individual and collective, and between public and private, as political philosophy has always done.

Third, at a 'micro' level, the politics of vulnerability can be understood as the ways by which power is related to the human body, the self, and the subject. We exercise power over our own body and self (we are subjects in an active sense), and others do so as well: institutions can discipline our body (we are subjects in a passive sense, i.e. we are *subjected*). This is what I take to be Foucault's main point in his early work (Foucault 1977). His analysis of disciplining also shows that we cannot say that vulnerability related to the self is a private and individual matter alone. Rather, our (modern) society promotes particular ways of handling the vulnerable body.

But conceptions of the relation between power and existential vulnerability should not be limited to what can be said within Foucault's body language. We can construct an account of power and politics by starting from the concept of existential vulnerability itself and *then* developing it by using Foucault. As I wrote in Chap. 3, vulnerability originates in, and is constituted by, the possibility of violence. Someone or something threatens me. The threat of violence influences me and shapes me as a subject. Exercising power does not necessarily mean brute enforcement, as Foucault knew (Foucault 1977). The shaping of the subject is a more subtle (but profound) kind of influence. If my subjectivity is shaped by something, this means that this something or someone has *power* over me. The vulnerability relation is at the same time a power relation. Struggling against vulnerability, then, is partly an attempt to (re)gain power and make one less dependent on another thing or another. A risk has a hold on me; when I experience that I am at risk, I am caught by it and I am in its chains. If I want to make myself less vulnerable, I try to *liberate* myself. However, if this power relation is understood in terms of a body that is made docile, as Foucault does in his earlier work, the power relation is a one-way relation: there is the one who exercises power and there is a victim. Moreover, Foucault's earlier work lacks sufficient attention to the active dimension of vulnerable existence. To avoid focusing exclusively on the victim and on the body, we should consider that vulnerability is not only shaped by that what threatens us but also by our own actions, by our standing-out and being-involved in the world. We are not merely an instrument that is being played; we also play and (in the process) rebuild the instrument. We are not just *subjected* as selves; we also subject ourselves. We are not mere passive bystanders of the miserable spectacle of our vulnerability; instead, we give meaning to it, we co-shape our vulnerability relations, and there is both suffering and joy in these relations. Using Foucault's later vocabulary, we shape these relations by means of 'technologies of the self' (Foucault 1988). However, this should not be limited to what we do with or to our body. We are subjects-in-relation, both (1) in relation to the world and (2) in relation to others. First, Foucaultians who exclusively focus on 'the body' and those who argue for bodily enhancement either have a non-relational, dualist, and Cartesian view of human being, or react against that view by overemphasising the body. Second, the shaping of vulnerability and the shaping of the self are also *social* matters, and this social dimension should not be understood only in terms of disciplining by modern social institutions. We cope with vulnerability together and as I said before, solidarity can also grow out of it.

Starting from this understanding of the existential vulnerability relation as a power relation, we can then turn to the issue of political power at 'macro' level. My vulnerability is partly constituted by the vulnerability that arises in the relation between me as a citizen and those who exercise power over me. Usually, the main political power takes the shape of a (nation) state. As I argued in Chap. 4, this gives rise to the threat of (state) violence. From a Hobbesian-contractarian point of view, this threat is necessary for social order. The Hobbesian idea is that the state has a monopoly on violence in order to prevent violence among citizens. But the state can misuse its power monopoly. When this happens, political vulnerability collapses into political violence—in particular state violence. In either case, the political order—for instance, in the form of the nation state—transforms our existential vulnerability. The threat of violence no longer only originates in others (social power) but also in the political power (here the state). For contractarians, this is regrettable but inevitable: we need a Leviathan to secure individual rights. However, there are also other, noncontractarian ways of thinking about politics and about how we can become political beings. For example, Aristotle thought that we are political beings by nature. Against Hobbesians, we can argue that there is no 'state of nature' (see also the previous chapter); there always was a *polis*, a political community, and we have always been political vulnerable in this sense. If we are political by nature, then to be politically vulnerable is as 'natural' as other vulnerabilities.

Note, however, that this Aristotelian argument does not imply that the political community has to take the form of a nation state. And some forms of political community make one more politically vulnerable than others. For example, the nation state seems to be a particularly risky form of political community since with its nationalist identity politics and its centralised form of political power, it creates the risk of state violence against citizens with different identities (who may challenge the legitimacy claim of the central government and may want to liberate *their* people and *their* nation from oppression) and the risk of violence between national states (nationalist wars). Moreover, some societies are more vulnerable than others, due to natural, social, and political factors (see also my remarks about the geography of vulnerability in Chap. 3).

Fourth, within a society, some people are more vulnerable than others. There is a vulnerability distribution. People are not 'born equal' when it comes to vulnerability, and upbringing and formal education maintains or changes the distribution—for example, it may widen the initial gap. Moreover, some people may have more power to mitigate and transform their vulnerability, for example, by means of financial resources, social networks, information technology, and (other) enhancement technologies.

Furthermore, *globalised* technology and economy makes some people more vulnerable than others. As Beck remarked, 'some people are more affected than others by the distribution and growth of risks, that is, social risk positions spring up' (Beck 1992, p. 23). Although some risks 'catch up with those who produce or profit from them' so that 'even the rich and powerful are not safe from them' (p. 37), there is still an unequal distribution of risks. We are all vulnerable and we are all mortal. But some are more *vulnerable* than others and indeed *made* more vulnerable than others by socio-economic arrangements. For example, when hazardous industries are transferred to developing countries, we are creating 'a global proletariat of the global risk society' (p. 41). To put it in Beck's economic vocabulary: there are producers and consumers of risk (p. 46), and although people might find themselves in both categories with regard to particular risks (global risks), often there are large differences in local risk and vulnerability. To use Marxist terms, we could say that there always have been, are, and will be social 'classes' defined by their degree of vulnerability as transformed by globalised technology and economy: *vulnerability classes*. There is a kind of 'vulnerability proletariat' (which may or may not revolt against its oppression). This is another sense in which there is a 'politics' of vulnerability and of technology. How vulnerable you are to technological risks and what you can do to manage, evade, or diminish the risk depends on your power position, which is at the same time a vulnerability position. We could describe and analyse these natural-social processes and structures. However, although such descriptive efforts will always have a normative component, we should also consider the possibility of a more explicit normative approach. We could think about why a particular vulnerability distribution is unjust and we could think about what we should *do* about that injustice. A normative politics of vulnerability could seek to redistribute vulnerability (see below), for example, by redistributing capabilities in order to mitigate and transform vulnerability. To use the warrior metaphor: such a politics would 'arm' people to struggle against their vulnerability. To use another metaphor: such a politics aims to give people the tools and the training to re-craft their vulnerability (see also my discussions of vulnerability style and vulnerability art in the next chapter).

Fifth, and related to the former point, a 'politics' of vulnerability does not only concern what could happen to me but also what could happen to *others* as a result of *my* anti-vulnerability strategies and actions (see also my previous remarks about vulnerability and violence). A macro-perspective that only focuses on 'distributions' risks losing sight of the dynamic and intersubjective dimension of coping with vulnerability. One way to lower my vulnerability (reducing power exercises over me) is to raise the vulnerability of someone else (by exercising power over the other). Some may hold that vulnerability is not a zero-sum game, and seek partners and co-warriors in the struggle against vulnerability. Others will certainly think it *is* a zero-sum game, see others as competitors, and act accordingly. Competition models can be institutionally and culturally spread and competition can be institutionally and culturally amplified, for example, in the context of Western individualist culture and economic competition. The zero-sum approach demands ethical-political constraints we should put on (anti-)vulnerability struggles (e.g. from a Hobbesian perspective), whereas the non-zero-sum approach is far less pessimistic and emphasises natural-social cooperation.

Sixth, in light of the discussion about human enhancement, the question regarding the politics of vulnerability can be understood as the ethical question posed at the collective level: if these new technologies are being developed and will have implications in the future, what stance do *we* take as a society? What stance should international and transnational organisations take? Who should decide about these issues? This leads us to the problem regarding the relation between ethics and politics, and between technology and politics—a familiar problem related to familiar political-theoretical questions. If there is one right ethical answer, why do we need majority democracy? And if there are many answers, how do we choose the best one? Is our current political system sufficiently equipped for responding to ethical questions posed by new and emerging technologies? Should we perpetuate our habit to locate political power in the political institutions and capitals of our nation states, or should we (further) globalise political power or search for power relations in the laboratories of scientists and companies, which are not necessarily located in the 'old' political centres? Should we limit participation in politics of technology to elections in a representative democracy, or should the public participate in decisions regarding the development and use of technology in other ways? Should we hold on to consensus models of decision-making or adopt an agonistic democracy model (e.g. Mouffe 2000) as the ideal model of the polis? We can try to imagine the future of our societies and ask how they should be organised taking into account the availability of human enhancement technology. And although the question is asked at the collective level and in a detached way, the answer will depend on how such technology will impact on concrete, personal lives: on our *existence*.

Finally, we might take a theoretically conservative position: the politics of vulnerability can be understood not as requiring the reinvention of politics, but as the application of our current political principles to imagined futures. If the nature of politics remains similar, then we can ask the usual questions of political philosophy: we can evaluate possible futures by using criteria such as freedom, equality, community, and justice and by using political theories such as utilitarianism and contract theory. For example, we may ask: How can we as a society organise human enhancement in a fair and democratic way? (See one of the next sections.)

In the next sections, however, I will *start* discussing the relation between politics, vulnerability, and new technologies by using existing principles and theories, but this discussion will also challenge us to think differently about the political. I have structured my discussion along the lines of the application of two important well-known political concepts: freedom and justice. But by focusing on technology and vulnerability, I will reinterpret these concepts in the light of my analysis of vulnerability. As a result, a less familiar picture of politics will emerge, leading to what can be framed as a preliminary exploration of a nonmodern conception of politics and its relation to vulnerability. For example, in the course of my discussion, it will become clear that too strict modern distinctions between spheres such as 'religion', 'politics', 'economy', and 'psychology' are untenable. Furthermore, the distinction between public and private will be questioned; it will be shown that the problem of vulnerability cuts through this modern distinction. It will turn out that discussing normative political questions regarding human enhancement, information technology, and vulnerability involves more than applying existing political-philosophical principles and theories: it requires us to revisit, reinterpret, and modify these principles and theories. This will help to define and develop the project of reinventing the political in the information age.

Note that although in the next sections I will focus on 'macro'-political issues, it is good to keep in mind that the ways we cope with each issue as a society have existential consequences: consequences for the way we experience and cope with our (personal) vulnerability. As the previous chapters have shown, transformations of vulnerability are at the same time natural, social, psychological, spiritual, and so on. For example, a financial system or a social institution has direct and indirect consequences for personal vulnerability. Thus, although I will start discussing 'political' and 'social' issues now, they are also personal-existential issues. As I will show, to deny this is itself problematic (e.g. if we separate 'private' vulnerability issues from 'public' risk issues).

I will also use this opportunity to further develop my analysis of *technological* transformations of vulnerability, in particular the ongoing transformations by new information and communication technologies (see also the previous chapter) and the 'human enhancement' transformations projected by the visions of transhumanists.

8.2 Freedom and Vulnerability

8.2.1 The Hobbesian Problem of Social Order: Neo-feudal Power Structures in Cyberspace

To many people, cyberspace appears as a realm of freedom and liberation. Some celebrate this freedom; others are afraid of the chaos and anarchy they perceive. Let me explain this.

'Liberation' can be understood in the metaphysical and transcendental sense I discussed before—transhumanist visions of leaving (behind) the body, for instance—but also in a more concrete way that is closer to the mundane meaning of 'politics': we can use information technology against oppression and censorship by less democratic and nondemocratic states (and we might want to add: against manipulation by private companies). Information technology can be an instrument of revolution. For example, social media played a significant role in recent Middle East protests and demonstrations (sometimes referred to as 'Arab Spring'). Giving examples of online protest in China, Diamond has called the Internet a form of 'liberation technology': it 'enables citizens to report news, expose wrongdoing, express opinions, mobilise protest, monitor elections, scrutinise government, deepen participation, and expand the horizons of freedom' (Diamond 2010, p. 70). And WikiLeaks facilitates mass document leaking in order to reveal the 'truth' about actions by states (e.g. in war). Furthermore, some see hacking as a means to reveal hidden information and to rebel against oppressive authority. For example, some people define themselves as 'cyberpunks' who hack as a way of contributing to less authoritarianism and more freedom, and (some of them) prefer anarchy.[1] Thus, it appears that new information technologies are liberation machines.

[1] See, for example, some definitions on http://www.ecn.org/settorecyb/txt/shortly.cypunk.html

But there are also other voices. Many people fear the freedom and (apparent) anarchy of the Internet. More freedom also means more vulnerability: for governments and companies, as well as for users and citizens. From this perspective, it seems that our security and privacy is at risk in a seemingly 'lawless' cyber world. What we may call the *World Wild Web* is a great environment if you are an ICT expert (e.g. a 'hacker'), but for less 'informationally' empowered people, it can appear as threatening and, to some, as hell. For example, Levmore and Nussbaum have recently argued in *Offensive Internet* (2010) that the Internet, while it benefits many people, tends to be offensive and nasty when it threatens our privacy, facilitates harassing and abusive speech, provides false information, and damages reputations and lives. Then, its lawlessness is a problem rather than a liberation, and several authors argue for legal remedies and all kinds of (other) regulation—including constraints to speech in order to control 'threats to privacy and reputation' (Levmore and Nussbaum 2010, p. 11). They understand the Internet—including 'Googling'—as being 'fraught with peril' (p. 2).

Although both positions (let me call them the optimists and the pessimists) differ in their moral evaluation of the freedom new information technologies seem to provide, they share the basic presupposition *that* the Internet and related technologies and media create a free, 'lawless' space where almost everything is allowed (and possible). It is assumed that the Internet is a space for *outlaws*, a space where one does not need to follow the norms and laws of society, indeed a place outside society. It is understood as a nonsocial, asocial, or pre-social space which needs to be kept hospitable by protecting its anarchy (optimists) or which is to be made hospitable by creating a decent society (pessimists).

In the political-philosophical tradition of liberal contractarianism, with which both optimists and pessimist are closely allied (not necessarily knowingly and willingly), this condition is called a 'state of nature'. Cyberspace is viewed as constituting a (new) state of nature, which then can be evaluated in two different ways. According to the optimistic, Romantic interpretation, which has roots in the work of Rousseau, the state of nature is a good place and people are naturally good. Society corrupts and states corrupt even more. They alienate us from our original nature. We should try to keep the purity and naturalness of the state of nature and of the human by all means in order to keep our original freedom. This Romantic view is what I take to shape the libertarian position (we might include cyberpunk here). Our vulnerability arises from the oppressive and corrupting social and political institutions we constructed. According to the pessimistic, Hobbesian interpretation, however, the state of nature is nasty and brutish. We risk a *bellum omnium contra omnes* (a war of all against all). Therefore, we need to create society and, in Hobbes's view, a strong if not totalitarian government, the monster *Leviathan*, which is a deplorable but necessary evil to make sure that people do not harm one another, since, in contrast to what Rousseau thought, people are *not* naturally good. This is what I take to be the position of those who defend a strong state. They fear the chaos of the Internet; (regulation by) the state is their anti-vulnerability strategy.

Of course, there are positions in between those extremes, in particular the position that we need *some* legislation and (other) regulation. This is the position liberal humanists like Nussbaum defend. But so far, all positions outlined here share the basic assumption that the state of nature is asocial and that cyberspace constitutes such a state of nature. The social and the political—good or evil—are created 'on top' of the bare asocial ground.

In practice, of course, we observe that cyberspace is no longer a 'pure' state of nature (and probably never has been; see the next section). Even if we think that there *was* a state of nature, we see neo-feudal power structures emerging. Hackers unite, companies and governments ally, individual users take security measures, and politicians roll out regulation. But these phenomena can be accommodated with the contractarian (and in particular Hobbesian) way of thinking. To mitigate your individual vulnerability, you cannot only use anti-vulnerability *shields*; in addition, you better choose to cooperate with others. In a cyber *gang,* you are stronger. There are cyber *warlords*. There are *castles* in which we can try to hide, for example, by using antivirus software provided by large players on the information and communication technology market. But it still seems that, on the whole, cyberspace does not (yet?) know stronger, fixed social institutions and a strong global 'cyber' state. The neo-feudal alliances are not stable. There is no global 'cyber' social contract—a global situation rather similar to the global 'offline' world—at least if we disregard some attempts at social and political institutionalisation by non-governmental movements and by, for example, the United Nations. The situation does not really change. Cyber libertarians can sleep in peace: they can dwell in (a sufficient amount of) liberty. Pessimists remain worried about what they see as a sphere of anarchy and crime.

These contractarian approaches to the social impact of new information technologies are attractive frameworks which we can use to make the (cyber) world a better place. We can take measures either to protect liberty (cyber libertarians, anarchists, etc.) or to restrict liberty in order to avoid harm. However, there is also an alternative way of thinking about (the politics of) cyberspace.

8.2.2 Alternative: Why Cyberspace Is Already Social

According to the alternative view I intend to develop here, both pessimists and optimists start from the wrong assumptions. Cyberspace is not a state of nature; *it is always already social*. We do not enter it as pure and free individuals; we enter it as social beings and bring our social vulnerabilities with us. Moreover, there is no cyberspace in the first place that is separate from the social. Our *onlives* are socially vulnerable. There is no sharp distinction between a sphere of (nearly) total freedom—understood as the absence of constraints—and a sphere of law, regulation, and civilisation. We already live in a social world with norms, rules, and customs. Of course, norms, rules, and customs are different in different environments, but these environments are not exclusively online or offline; rather, they are 'hybrid' environments, mixtures of online and offline. The relevant social norms and the games and practices to which they belong cut through this distinction. For example, communication between colleagues (online and offline) is different from communication between friends or between lovers (online and offline).

However, recognising this non-dualist (social) ontology does not imply that, for example, the Internet is neutral with regard to the precise form the social takes. As I argued before, the Internet—and, more generally, new information and communication technologies—transforms our vulnerabilities, including our social vulnerabilities, in various ways. We must take seriously the violations reported by Nussbaum and others, for sure, but we should do so for the right kind of reasons. We should not understand these violations as part of a war of all against all that goes on in a separate cyberspace, a universe without rules where we are all *outlawed* and where only barbarians without refinement live. The distinction between online and offline—if it makes sense at all—should not be used to separate the wild from the civilised. Liberal humanists such as Nussbaum who see 'the Internet' as oppressive tend to presuppose that on the one hand, there is a secure and civilised offline world where we have the technology of *books* to cultivate people and where we are safe (the classical humanist technology), and on the other hand a perilous online world where new information technologies create a breeding ground for nasty speech and criminal acts. They suppose that what should be done is to finally and properly colonise that threatening, wild territory in order to control and civilise it. To put it simplistically, the recommendation is establish an Internet police (for Hobbesians, you can only have a *polis* if you have a strong state with a strong police force) and teach people how to read books, which will make them into better people (the humanist solution).

But this position is blind to the transformations of society, of humanism, and indeed of 'humanity' that have already taken place and that are still taking place. The new technologies increasingly co-constitute what it means to be a social and political being: 'engaging with the Internet' or living 'in' cyberspace is becoming part of what it is to be such a being. The Internet does not exist outside society; there is *one* social reality, which is shaped and transformed by the current information and communication technologies. This does not mean that there are no ethical and political problems. An alternative interpretation of the phenomena Nussbaum and others rightly worry about is to say that new information technologies indeed transform our *onlives* and hence our social vulnerability. The world becomes our village, the schoolyard is expanded to the world, and this has consequences: our social vulnerability is also globalised. If there is a global village and if there is no longer a distinction between private and public, then the 'they' that can do something to me can be *everywhere.* The new environment mediated by the Internet allows those who threaten us to remain anonymous and invisible. This *does* increase our social vulnerability. But the sources of that vulnerability are not restricted to 'the Internet' if that term is meant to refer to a separate space. There is one social space which is at the same time online and offline. We do not have on the one hand a 'virtual' online identity and on the other hand a 'real' offline identity; the virtual and the real mix. We are what we are as social and technological beings. We have one reputation and one life.

Moreover, what also makes possible the violations and the wounds under consideration here is not so much *too much* liberty in a separate sphere, but a highly *problematic understanding* of liberty and of our relation to others, which influences how we perceive and treat others—anywhere. The individualist-contractarian view itself is problematic: it is only by presupposing that we are individuals isolated ontologically from others who have inalienable rights and absolute liberties (e.g. freedom of speech), that we can even consider not only claiming these rights but also *exercising those liberties in an absolute way*. If we instead understood ourselves as social beings whose liberty is deeply dependent on others, on the social, then we may be much less inclined to talk and behave in harmful ways towards others. (The same is true for our relation to the natural environment, which I will discuss later.) Thus, the problem is that because we see ourselves as asocial, we create the conditions that promote behaving *anti*social. Furthermore, since in this modern way of thinking we understand ourselves as individuals and attach great importance to our ego, we are easily offended when others criticise us. In this way, we render ourselves more vulnerable to the social risks associated with globalised *onlife*. If we see ourselves as islands (we think we are an individual) in a sea of hate and violence (the social as it appears to us), we are more likely to act antisocially and we are more likely to suffer from antisocial acts, from actual violations and from what I have called second-order vulnerability. Indeed, we also suffer from the awareness that our privacy, identity, and reputation can always be damaged by others. The potential victim is also a victim.

What can be done then to decrease social vulnerability co-shaped by new information technologies? If the social is globalised and if the distinction between online and offline is weakened, then at times when the social takes on the appearance of the 'they' that threaten me (but not at other times), it seems that there is no exit. We cannot retreat to the private sphere (since the private is now public) and we cannot retreat to the 'real' offline world (since there is only *onlife*). The victim seems to be locked up in a totality from which there is no escape. But if we do not want to (and perhaps cannot) return to a world without new information technologies and media and if we cannot and do not want to return to *local* worlds, what is the alternative? Is there a way out?

One solution is to regulate speech and take other measures that decrease freedom. This is what Nussbaum and others seem to propose. For example, Brian Leiter criticises what he calls 'the rhetoric about "free speech" in cyberspace' and suggests that we regulate some kinds of offensive speech on the Internet (Leiter 2010, p. 172). But this would only create a new kind of vulnerability—(more) control by the state—and might even generate a new kind of *totality* from which there is no exit: a totalitarian state that controls 'online' and 'offline' environments. How else could we control social environments in such a way that all abusive and hateful speech is avoided? But this solution would create a situation most of us consider worse than the present situation, since it multiplies the 'they' and gives one of them absolute power: the state, Big Brother not only watches you but also listens to what you say. The cost of this solution in terms of freedom is too high. (Note that, in so far as this solution has already been chosen—in the West and elsewhere—we are already paying some of that price.)

Another solution is to work on strengthening the social fabric itself. One problem with the picture of social vulnerability I just sketched is that it seems to presuppose that we stand alone, that there is no-one to support us at times when we are the victim of the 'free speech' of others, and that there was no education and no experience (and wisdom) that prepared us for such violations. The modern way of thinking about the social—which of course started long before the Internet—brings about a world that increasingly resembles that nightmare scenario. But there is an alternative possible (social) world, which might develop if we were to start thinking differently about the social and our relation to it and take institutional measures that, instead of prohibiting particular acts and speech, promote the growth of social environments and social networks that help us to decrease our social vulnerability and are able to support us when we need it. Then, social networks become better *safety nets* and help us to better cope with social risk.

These networks neither belong to the 'offline' world nor to the 'online' world; with regard to what they do, this distinction does no longer matter. Acknowledging the new shape of the social means to demand that we be *socialised* into beings that live in today's world, with its new information and communication technologies. They require that we educate our children into beings who are less vulnerable in the global spaces (which are online and offline at the same time). They require that we set up social institutions that decrease social vulnerability and help the most vulnerable among us to cope with their vulnerability. Again, such institutions do not belong to either the 'offline' world or to 'cyberspace'. They belong to one world, our world.

This suggestion of a more nonmodern way of thinking and doing with regard to the social does neither imply (1) that we can easily 'switch' to that alternative way of thinking and doing nor (2) that in the meantime we should do nothing against the abuses and the threats that arise in the social world as it is perceived by those who think in a modern way. First, we live in a society and a culture that facilitates modern ways of thinking and doing. This renders it harder to imagine and act in a different way. We have no reason to expect that profound changes to a culture come fast—if they come at all. Second, even if most people still perceive 'cyberspace' as a separate sphere, this should not keep us from discussing solutions within that kind of thinking. Within modern thinking *and* within the liberal-philosophical tradition broadly conceived, we can find and use other conceptual tools that can mitigate social and other vulnerabilities in their various forms. Even those who call themselves liberals do not hold that freedom is the *only* moral and political value. In the next sections I will (further) discuss the concepts of *justice*, *freedom,* and *democracy* in relation to vulnerability and technology.

8.3 Justice and Vulnerability: The Case of Human Enhancement

As I wrote in the first section: all people are vulnerable, but some people are more vulnerable than others. This observation can be framed as raising an issue of *justice* and fairness. Is it just that some people are more vulnerable to climate change than others? Is it fair that some receive a better education and are therefore less socially vulnerable than others? Is it just that some people have better access to health education and advanced medical technologies and hence can significantly reduce their bodily vulnerability, whereas others die if they have a serious illness and have to watch some of their children die from diseases which elsewhere in the world are regarded as historical? Is it just that some people are poorly educated and are extremely vulnerable to unemployment and poverty, whereas other people are born in countries where many good jobs are available, where all people have access to higher education, and where there is a decent welfare system that shields everyone from (extreme) poverty when all goes wrong in spite of all these benefits and all that education? And is it fair that some people are loved and are successful (partly) since they are born intelligent, beautiful, healthy, and well-cared for, whereas others have a much worse rather vulnerable 'starting position' and have to struggle to get what others take for granted, if they ever get any of it at all?

If these are problems of justice at all (see below), then how can politics respond to it? The standard answer is to create and maintain institutional arrangements that provide sufficient education, health care, and so on for all—or certainly for the least advantaged. Thus, governments usually provide compensation. It can also take pro-active measures to avoid injustices. A more radical and highly controversial way to create more 'vulnerability justice' in society, however, might be to use human enhancement technologies: they may enable us to create a different distribution of 'genes' (see below). But human enhancement can also be seen as part of the problem: it could create a society of 'haves' and 'have nots', or at least further widen existing gaps between 'haves' and 'have nots'. In this section, I will discuss the issue of justice in relation to human enhancement. Would the use of enhancement technologies as proposed by transhumanists create an unjust society or make it more unjust, and if so, what could be done about it? First, I discuss the Rawlsian question regarding justice as fairness (in particular *distributive* justice) in relation to the possibility of genetic enhancement. I also ask what the problem is and *what* should be distributed, if anything. In particular, I question the idea of a redistribution of 'genes'. Then, I turn to the capability approach in order to further develop my discussion of the relation between justice, enhancement, and vulnerability.

8.3.1 Towards a Fair Distribution of Genes or Vulnerability?

8.3.1.1 What Is the Problem? Is There a Problem?

On the one hand, in the absence of political intervention, human enhancement as proposed by transhumanists is likely to widen the gap between the advantaged and the disadvantaged. For instance, some individuals will be able to benefit from genetic enhancement (one form of human enhancement); others will not have access to these technologies. Perhaps, they will have been born in the 'wrong' place, have the 'wrong' nationality, lack sufficient financial resources, or will not have access to the correct information. But is this a problem of distributive justice, and why? And if it is one, is it really different from existing problems of distributive justice? Is it a *new* problem?

On the other hand, human enhancement can be viewed as a means to *solve* rather than cause problems of distributive justice. On this view, principles of distributive justice do not rule out, but instead justify or even *demand* human enhancement in order to improve the position of those who are less well-off—genetically and otherwise. If this is right, then we must ask what should be distributed according to what principle of distributive justice. For instance, Farrelly has argued that we should apply a 'Genetic Difference Principle' to genetic potentials (Farrelly 2004, 2005). But is this desirable? Is it even *possible* and *intelligible*?

Let me start with enhancement as a problem of justice. To say that enhancement *creates* a problem of distributive justice assumes that enhancement—of whatever nature—gives a comparative social advantage to those who benefit from it, and that, as a matter of fact, it is likely that if the technologies are introduced, some will actually benefit from it while others will not. An additional worry with regard to distributive justice and human enhancement is not only that at a given time some individuals will be enhanced whereas others will remain unenhanced, but that this gap between enhanced and unenhanced will be maintained or become wider in the future if inheritable traits are changed by genetic enhancement. It seems that we would witness the rise of a 'class' of people who are genetically enhanced and enjoy the benefits of this enhancement on top of the other benefits they already have (benefits which gave them access to the technology in the first place). What first seemed to be a 'biological' gap becomes a *social* gap—not only in the first generation but also in the next generations. If enhancement equalised social positions, there would be no problem of *distributive justice*, since the question of justice is asked on the basis of a comparison of relative positions (this would not exclude the possibility of *other* problems). But this option is unlikely to be realised; it seems very likely that human enhancement would create a social gap, or rather worsen the existing social gap.

But why exactly is the gap between advantaged and disadvantaged a problem of (distributive) justice? Before moving on to my main discussion, let me distinguish between four initial ways to respond to the prospect of human enhancement in relation to distributive justice.

One option is to hold that the current distribution is just or to deny that there is a problem concerning *justice*, and limit political intervention to a minimum. For instance, those who do not see a problem (of justice) with the current disparities in income and well-being are not likely to worry much about future large distributive gaps as a result of genetic enhancement technology. At most, they will defend some legal measures to ensure an open and competitive market. Various philosophical justifications have been provided for this view. For example, Hayek famously argued that using the term social justice is a 'category mistake': 'Strictly speaking, only human conduct can be called just or unjust. (…) A bare fact, or a state of affairs which nobody can change, may be good or bad, but not just or unjust.' (Hayek 1976, p. 31) According to him, this does not only imply that nature 'can be neither just or unjust', but also that society and the market, in so far as they are (or create) a 'spontaneous order' (p. 32), are neither just nor unjust. However, whether or not this view is right, when it comes to human enhancement, it seems that society *and* 'nature' become a lot less 'spontaneous' than before, since they become the product of human actions and hence raise the question of moral responsibility. If the transhumanists are right, we *would* be able to change the state of affairs, and Hayek's argument would not apply.

Another response is to agree that there may be a distributive justice problem in the future, but to focus on current distributive injustice rather than worrying about what may happen in a future we do not know. For instance, Lindsay has argued that we should focus on lessening 'the glaring disparities of the present' (Lindsay 2005, p. 33)—although his position is more complex than this quote suggests. However, this view disregards that there is often a thin line between therapy and enhancement and that in this sense enhancement is already going on. Moreover, worrying about the future of humanity does not exclude worrying about the present condition.

A third approach is to take seriously the problem of genetic enhancement (as a problem of justice) and its relevance for contemporary morality and politics while defending a ban on genetic enhancement to avoid the future problem. In this view, one of the reasons why we should stop developments towards genetic enhancement is that they will most certainly lead to a wide gap between the advantaged and the disadvantaged and—so it is assumed—this is unjust by definition. Whatever other (moral and political) reasons there may be to stop or stimulate the development of genetic enhancement technology, in this view, the distributive justice problem is a sufficient reason to ban it. However, this view assumes that we cannot do anything about the problem, that we cannot ethically guide and politically regulate human enhancement in order to avoid or at least diminish the gap that may be created.

A fourth possibility, then, is to acknowledge the distributive justice problem, but argue that we can both have genetic enhancement technology *and* undertake political action now and in the future to ensure that the distribution of the benefits and burdens of this technology is *just*.

Thus, the first view assumes that the current distribution is just (or neither just nor unjust), the second view assumes that the current distribution is unjust, the third assumes that the future distribution will be unjust, and the fourth avoids making these assumptions and allows for the possibility that the future distribution *may* be *made* just. But what is a just distribution? This question takes us to theories of justice.

8.3.1.2 Principles of Justice: Rawls

Let us start with a very well-known and influential theory of justice, which fits within the liberal-contractarian way of thinking: Rawls's theory of justice (Rawls 1971). The liberal-contractarian problem is how to maintain social order and cooperation, and principles of justice are a response to that problem. But what are the principles of justice? How can we *know* them?

Rawls's conception of justice as fairness gives a procedural answer to the question. The thought experiment of the original position is meant to generate principles of justice. Influenced by Kant, Rawls rephrases the question regarding principles as follows: what kind of distribution would rational persons hold fair, making abstraction from their personal interests and desires related to their social position? In the original position, persons are ignorant about their social position of society; they choose under a 'veil of ignorance' (Rawls 1971). It is reasonable, according to Rawls, to choose (as *one* of the principles) a principle that allows more general well-being only on the condition that the position of the worst off is maximised (the so-called difference principle). This proposal can be seen as a reaction to utilitarianism, which only considers consequences for the average position, not for the position of particular categories of people.

Suppose that this thought experiment is the best tool to evaluate justice and that the 'difference principle' is indeed the best or most likely outcome of the thought experiment. Then, we may want to apply it to the case of human enhancement. But let us first further unpack the problem. It seems that the problem concerns not only differences in social status. The worry is also that human enhancement may give rise to a situation where discrimination on the basis of *genetic* position is itself considered as a problem—rather than *social* position and its relation to a given genetic situation. The idea is that whereas usually the 'natural' distribution of genes is not seen as a problem of justice, a new problem emerges when the human genome is no longer a 'given' but becomes material that can be manipulated. In the usual thinking about justice (as in the discussion about Rawls), genetics is an important factor in influencing your social position, but is *not itself an issue*. The question of justice is, among other things, about how to *cope* with genetic variety in the human species, but the distribution of genetic 'capital' is taken as a given. With the availability of human enhancement, this changes in a significant way. It is possible that some people get enhanced and others not, which opens up the possibility for unfairness.

Thus, there are two possible questions of justice and indeed two possible positions with regard to human enhancement and justice: one asks 'What is a just distribution of genes' and the other asks 'What is the just distribution of social positions *given* that distribution of genes'. The second question we are already faced with, and if we ask this question, we can already use the theories of justice that are available to us. The social question today is already partly about the relation we want between 'natural' talents, on the one hand, and social position, on the other hand. Human enhancement may also raise very well-known questions with regard to discrimination. 'Genetic' discrimination can be explicit but also implicit. Even if your employer does not take a blood sample to check your genetic code when you apply for a job (as is shown in the film *Gattaca*, for instance), and therefore does not *directly* discriminate you, it is likely that he will favour someone who has better genes for the job on the basis of other tests that are strongly influenced by genetic factors. In such a future, although the own effort of people will still count, on average, people with better genes will be more advantaged—and indeed they are *already* more advantaged today. But human enhancement would also create an additional, new problem: it would widen the *genetic* gap between people. Then, a new relation would be in place: one between human action (enhancement) and the no longer 'natural' distribution of genes. *Then, the genetic distribution itself would no longer be taken for granted.* This raises the first question of justice: What is a just and fair distribution of genes?

A Rawlsian-style thought experiment helps us to respond to this problem. In the original position, there is ignorance about one's genetic make-up. You do not know if you are enhanced or not. In this situation, a Rawlsian might say, reasonable people will choose a society that is organised in such a way that genetic enhancement is allowed to benefit society only on the condition that (next to other requirements) the genetic position of the least advantaged is maximised. In other words, the difference principle is applied not only to the distribution of social position but also to the distribution of *genetic* position (I will criticise this application of the difference principle below, but let me first further develop this view).

The difference principle is an example of what is sometimes referred to as a 'prioritarian' approach to justice as opposed to a 'sufficitarian' one. *Sufficitarians* do not care about the width of the gap between the enhanced and the unenhanced, but are concerned that everyone passes a certain minimum threshold (Farrelly 2004, p. 23). For sufficitarians, human enhancement technology and policy would need to ensure that everyone has 'a *genetic decent minimum*', as Farrelly puts it (Farrelly 2004, p. 24). In a just society, there would be a guaranteed 'genetic minimum'. Achieving this aim requires enhancements for some who not have that minimum. However, this means that a gap between minimum and maximum persists, perhaps even a wide gap. Now such gaps are not *necessarily* unjust, but their (in)justice is not addressed by this sufficitarian principle. Sufficitarianism sets a threshold, but does not address the *relative* differences between people. Moreover, what is held to be a 'genetic decent minimum' will change over time; the content of any sufficitarian principle heavily depends on such changes.

The *prioritarian* approach provides a different perspective. Prioritarians think it is important to benefit those people who are worse off (Farrelly 2004, pp. 22–23). By giving priority to the situation of the less advantaged, as Rawls's difference principle does, the prioritarian principle seems close to that of the sufficitarians. However, in contrast to sufficitarians, prioritarians explicitly consider the *relation* between what happens to the disadvantaged and what happens to the advantaged. A sufficitarian politics also requires a 'genetic transfer' from the advantaged to the disadvantaged, but its justification is *external* to the distribution, to the gap between the advantaged and the disadvantaged. The difference principle provides an *internal* reason. It says that the distribution is not unjust if it meets the following criterion: the advantaged are allowed to get better off *if and only if* the position of the *worst off* is maximised. This is a formal principle, which is independent of any formulation of a particular threshold. Moreover, sufficitarians need not care about the position of the advantaged, as long as everyone has a certain minimum. A prioritarian considers the relation between what happens to the two groups. Both groups are acknowledged to be interdependent. Prioritarianism, therefore, fits better with the intuition that distributive justice is concerned with the *relation* between disadvantaged and advantaged. (However, this argument does not imply that this is a sufficient reason to adopt Rawlsian prioritarianism. I will discuss another, sufficitarian theory of justice below, which has its own attractions.)

This Rawlsian solution to the two problems of justice is different from at least four other ways of organising society in response to the possibility of human enhancement. First, one could reserve the enhancement to an elite, who are seen as fit to rule the others. The argument for the latter claim is that only the (genetically) best should rule. This is the Platonic solution. (Note that apart from genes, merit may still (be made to) play a role as a selection criterion as well in such a society: being 'the best' does not depend on genes alone but also on one's own effort.) Second, one could equalise all 'genetic potential'. This seems to amount to a kind of radically egalitarian and totalitarian solution, in which *social* position is equalised as well, but this option does not necessarily follow from the principle of genetic equality. Since one's own effort still plays a role, one could have a meritocracy as well—although a policy amounting to totalitarianism would probably be required to realise *genetic* equality. Third, one could leave the distribution of genetic 'capital' to the market. This solution accords with the 'liberal eugenetics' Agar has proposed, or with a kind of 'laissez-faire' human enhancement. It would probably go hand in hand with economic liberalism, or with laissez-faire economics. In Western societies shaped by that kind of political and economic thinking, it would be the most likely route if politics did not intervene in the technological developments that contribute towards radical human enhancement. Fourth, the political power could make sure that the *average* genetic position of people in (a particular) society is raised. Similarly, one could also try to raise the average position of 'humanity'—regardless of the positions of particular people. I take this to be the ideal of those transhumanists who are more occupied with the genetic progress of humanity as such than with that of a small group, a particular society, or themselves. The Rawlsian solution I explored, finally, rejects the totalitarian options, but allows state intervention (to avoid the disadvantages of laissez-faire and utilitarian solutions) in accordance with the difference principle and other principles provided by the Rawlsian theory of justice as fairness.[2]

But how non-totalitarian is this application of Rawls's difference principle really? It may not imply equalisation of genetic potential ('genetic equality'), but nevertheless there seems to be a conflict between the value of freedom and the value of justice. The objection from freedom runs as follows. The application of a prioritarian or sufficitarian principle to genetic potential would require compulsory genetic monitoring and 'treatment' of the 'genetically disadvantaged'. Now if we value negative freedom (freedom from interference by other people), *and* if we value negative freedom more than justice, we are entirely justified to reject

[2] Note that this Rawlsian excursion starts from the assumption that further developments towards human enhancement cannot be stopped. I am convinced by the argument (frequently made by transhumanists) that it is unlikely that a ban on (certain forms of) human enhancement will be able to prevent some people from developing it and using it, and that therefore it is better to think about how to organise a society in which human enhancement plays a role than waste our efforts on trying to stop it. Thus, a *total* ban on human enhancement may not be fully effective, whatever ethical position we wish to defend. Having said that, however, a Rawlsian solution does not exclude a ban on the *use* of *certain forms* of human enhancement, but may even require it. For example, it seems to require stopping an elite from reserving a certain form of enhancement to themselves only—although, of course, this ban too may not be very effective. But tax systems are also not always entirely effective, and still we use them and justify this by appealing to social justice.

the proposal for that reason. We do not want to go down the 'totalitarian slope'. The value of negative freedom is widely accepted, at least in the West. Moreover, there is a political consensus in Western nations that consent is crucial, and international organisations influenced by the Western discourse of freedom and human rights have adopted policies and declarations in which freedom is a *central* value.[3] Therefore, we should not locate freedom entirely outside the theory of justice; most influential theories of justice include (negative) freedom within their principles, including Rawls's theory. Nevertheless, even if we incorporate the value of freedom and make it part of a theory of justice in relation to human enhancement, then within that theory a tension between *freedom*-as-part-of-justice and *distributive* justice remains. If we want a politics of human enhancement that redistributes, therefore, we should carefully discuss its implications for freedom. We have to decide how much enforcement and restriction is acceptable; we have to balance freedom and justice.

But if the theory of justice we use requires redistribution, *what* are we going to redistribute?

8.3.1.3 What Is To Be (Re-)Distributed?

Indeed, a further problem concerns the object of distribution. If we want justice with regard to vulnerability and want to propose redistribution as a way to realise this, then what exactly do we want to redistribute? So far, I discussed the politics of human enhancement in terms of the question regarding the fairness of the distribution of *genes*. But this is problematic in several ways.

First, as I already noted in relation to genetic equality, one of the problems with such an approach is that since your position in society does not only depend on the genes you have, a fair distribution of genes is neither necessary nor sufficient for a fair and just society. It is not necessary, since adjustments to positions in society can be made by other means. How sufficient one thinks it is depends partly on one's view of 'merit'. 'Merit' refers to what you deserve based on your own efforts. But how much your 'own' are your 'own efforts', given that the person you are and what you can do in life, crucially depends on others and on what happens *not* as the result of your own acts (e.g. on luck)?

But there is also another problem. The relation between 'natural' genes and social position—and indeed between genes and vulnerability—is not very straightforward. What grows on the genetic 'platform' depends on more than 'genes' alone. This brings me to a more fundamental problem with the idea of a (re)distribution of genes or genetic potential. What is the 'genetic potential' I mentioned? What is the relation between genes and social position? Why change 'genes' if what we really

[3] For instance, UNESCO's Universal Declaration on the Human Genome and Human Rights says that 'prior, free and informed consent of the person shall be obtained' in all cases of research, treatment, and diagnosis (Art. 5b).

want is a different distribution of social positions? Is the metaphor of a 'platform' appropriate? We need to have an adequate understanding of what we are talking about when we talk about 'genes' in relation to the human and to social position. Let us turn to what contemporary philosophy and (biological) science tell us about 'genes' and their relation to what we are.

First, from an existential-phenomenological point of view, biology is only one, one-sided way of seeing the human. If we construct social position or our human identity in terms of 'biology' or 'genes', this is not the single true 'objective' definition of social position, but one particular way of understanding the concept, which unnecessarily and wrongly leaves out other dimensions of human being and of the social—and indeed other ways of perceiving and relating to our body and to our environment. To see ourselves as vehicles and expressions of genes (alone) does little justice to the variety and richness of human existential experience.

Second, even on a 'purely biological view', we are so much more than 'genes'. As Lewontin has argued, biology shows that there is a complex interplay between genes, organism, and environment (Lewontin 2001); biological traits are the result of genes, chance, and environment. The metaphor of a 'platform' or a 'blueprint' does not work. Moss has argued that 'genecentric' views, which place genes in central control of the organism's development, must be replaced with a decentralised approach that includes intercellular, biochemical, and sociological factors (Moss 2003). And Salvi has shown that it is unrealistic to suppose that we can manipulate human germ cells in a pre-ordinate way since the long-term consequences of such interventions are unpredictable (Salvi 2002, p. 74). Furthermore, if we cannot predict the phenotypic expression of bioengineered genes, then we cannot know what it will do to the individual—including whether or not it will 'enhance' that individual. If this is true, then transhumanists who suggest otherwise have unrealistic proposals. Moreover, it also means that it is problematic to speak of 'genetic' enhancement: it suggests a simplistic view of what we better call 'natural goods' (rather than 'genes') and their relation to social position.

Third, how much sense does the term 'natural goods' make? If the influence of the environment on our traits is as important as Lewontin and others say it is, then our social and cultural arrangements also have an influence on our traits and (hence) on our social position. Moreover, from a nonmodern philosophical point of view, the meaning of 'environment' should not be restricted to the 'natural'—as if such a space purified of the social existed. As Latour has argued, such strict modern distinctions are untenable (Latour 1993). Like many other hard moral and political problems, the issue concerning the relation between distributive justice and human enhancement is a 'hybrid' problem in the sense that it is concerned with complex entanglements of the natural and the social. At the very least, we should therefore reframe the problem of (re)distribution with regard to human enhancement as a problem of (re)distributing natural-social goods. If we adopt a hybrid natural-social ontology, then we can no longer retain a compartmentalised view of goods. Therefore, we must adapt our assumptions about natural and social goods in our theories of justice as well. The object is neither the distribution of social goods, as in existing theories of justice, nor the distribution of natural goods, but the distribution of natural-social goods.[4]

Moreover, if the relations between genes, organisms, and environments are indeed as complex as claimed above, and if there are equally complex dynamics and entanglements between the natural and the social, then it is plausible that we currently lack sufficient knowledge of how to achieve the goals of distributive justice by using 'genetic' and other human enhancement technologies and policies. At present (and in the foreseeable, near future), both our knowledge of genetics and our knowledge of the interplay between the natural and the social are too limited to achieve these purposes. If this is right, then it implies that human enhancement, broadly understood in terms of comparative advantage, cannot be properly and comprehensively justified or evaluated by referring to distributive justice as long as we lack sufficient knowledge about what enhancement may do to social position.

Given these problems, we might want to turn away from discussions about 'genetic enhancement' and the redistribution of 'genes'. In general, we might want to turn away from discussions about the distribution of any resource or goods and instead focus on the distribution of social positions itself. We could talk about 'output' or 'outcome' fairness rather than 'input' or 'initial' fairness. But what does this mean?

Taking seriously the points I made about vulnerability and politics earlier in this chapter, we could take *vulnerability* itself as the outcome variable. Consider again my observation that some people are more vulnerable than others. If we understand this as a problem of justice and fairness, in particular a problem of *distributive* justice, then a normative politics of vulnerability must seek to redistribute vulnerability. The idea of the original position, then, would be to discuss which principles should guide the fair *distribution of vulnerability* in a society. Vulnerability is then understood as the outcome of various factors, such as genetic make-up and biological traits (whatever that means), social position of the parents, education, your own efforts, and your current social position.

A Rawlsian-style thought experiment, then, would imply that we blind ourselves to these factors—that is, to our social and 'genetic' identity—and then think about which vulnerability distribution would be fair. If it is plausible that under such a

[4] Note that this also applies to capabilities, in so far as they can be understood as 'goods' at all. They are also both natural and social. Given my remarks on natural and social goods, we can no longer understand capabilities as pure 'social' or 'non-natural' goods or divide them up into 'natural' and 'social' goods. Consider Nussbaum's capabilities list: life, bodily health, senses, imagination and thought, emotions, practical reason, affiliation, being able to live with other species, play, and control over one's environment (Nussbaum 2006, pp. 76–77). Each of these capabilities can be considered natural-social capabilities. For instance, life and bodily health seem 'natural' goods, but cannot be considered apart from the technological culture and social institutions that promote or threaten these goods, such as medical care or crime. And thought and practical reason cannot be considered entirely separate from 'natural' bodies, from 'social' others, from society, and from the 'natural-social' environment to which they are directed. A just distribution of these goods, then, must concern the distribution of these 'hybrid' capabilities. However, in the next section, I will argue that capabilities are better interpreted as having to do with *outcome* rather than resources.

'veil of ignorance' the difference principle would be (one of the principles) chosen, then the political challenge is to make sure that those who are worst off in terms of vulnerability (the most vulnerable) benefit most from lowering the overall level of human existential vulnerability (e.g. by means of genetic enhancement). Thus, changing the object of distribution from 'genes' to 'vulnerability' seems to be an attractive and elegant solution to the problems indicated.

However, how is this 'vulnerability' to be understood for practical purposes, for example, when it comes to policy? Policymakers may wish that existential vulnerability could be clearly identified and perhaps even *measured*, but this is problematic given what I have said about existential vulnerability. One could take into account the 'factors' mentioned (genetic make-up, education, etc.) and construct a 'model' of human vulnerability, but *existential vulnerability* is not a matter of 'factors' or 'facts' and cannot be 'modelled'. As a relation, it has both objective and subjective dimensions. Threats of violence are born out of my relationship with the world. A description of that relationship is not exhausted by listing facts and figures, or by constructing a vulnerability model that takes into account various 'factors', but also needs sensitivity to the meanings that emerge in subjective experience and to engage in the interpretation of personal and cultural narrative. We need not only Rawls and Sen, but Jackson too. We need principles, but we also need hermeneutics. This 'narrative' approach does not mean fantasy. We may want to sharpen our ethical and political sensitivities by reading novels, as Nussbaum proposes, but fiction cannot be isolated from non-fiction (in this sense, there is no such thing as 'pure fantasy'): the analysis of vulnerability with regard to the issue of fairness needs a focus on concrete, daily experience of real people and real lives. As philosophers, we can learn from empirical anthropologists and the people they *talk* to, *experience*, and *interpret* (not 'observe' their 'behaviour'), journalists and readers, social workers and teachers, politicians and citizens, criminals and victims, and writers and poets.[5] And last but not least, we can learn from our own experience—as long as we do not forget that this too is not 'raw' or 'pure' experience but is always already interpreted and reflected on, formed and transformed. Writing is a technology of the self that does precisely that: it enables us to interpret and reflect on our experience and at the same time it shapes that experience.

To conclude this (sub)section: to demand justice with regard to vulnerability and human enhancement is problematic in various ways, especially if we value freedom and if we consider the idea of redistributing 'genes'. I proposed to turn from 'genes' to 'vulnerability' as the central concept, which is outcome-oriented and, if interpreted in the way I proposed, does more justice to the existential and experiential dimension of the human—thereby allowing us to reconnect with the approach I have been developing.

Does this mean that we must turn away from standard political-philosophical theory? Not necessarily; there is another way of rendering 'vulnerability' less

[5] I believe this is also the direction Nussbaum takes in recent work on law and capabilities. I will discuss Nussbaum in the next section.

abstract for the purpose of a political of vulnerability as a politics of justice: we could relate vulnerability to capabilities and use the *capability approach* as a theory of justice in order to evaluate vulnerability transformations by new technologies and to evaluate the idea of human enhancement. The capability approach is outcome-oriented, which means that we do not have to focus on 'genes' or other 'resources'. A further advantage is that capability theory has a 'built-in' concern for freedom: in Sen's version, for instance, it is about enhancing freedoms. However, using the capability approach within the framework I sketched (the project of evaluating existential vulnerability transformations and of evaluating the visions of human enhancement and of technology related to it) will also require its reinterpretation and modification. For instance, we will have to call into question the distinction between resources and outcomes, between ends and means, which is assumed by Rawlsians as well as by capability theorists (recently called 'capabilitarians'). This will be my programme in the next section.

8.3.2 Capabilities and Vulnerability Transformations: Human Enhancement and Human Development

8.3.2.1 The Capability Approach, Information Technologies, and Human Enhancement

The capability approach (CA) was developed by Sen and others in response to standard welfare economics and development studies, which measured well-being only in terms of material resources. Sen argued that what matters is how people transform these resources into real freedoms and valuable activities. Development should not be identified with the growth of the national product (Sen 1999) but with *capabilities*: what people effectively can do with their resources. Interestingly, capabilities are not just abilities (something 'internal' or 'personal'): they are also 'the freedoms or opportunities created by a combination of personal abilities and the political, social, and economic environment' (Nussbaum 2011, p. 20).

In Nussbaum's version of the CA, especially in her more recent work (Nussbaum 2006, 2011), the CA is elaborated into a philosophical theory of justice. It is based on the notion of human dignity (Nussbaum 2006, p. 74) and on the neo-Aristotelian insight that 'political animals' are not only rational but also social, bodily, needy, and vulnerable beings. Arguing from this philosophical basis, Nussbaum has listed a range of 'central' human capabilities, the fostering of which is a matter of social *justice*: to enable people 'to pursue a dignified and minimally flourishing life', a 'decent political order must secure to all citizens at least a threshold level' of the following capabilities (Nussbaum 2011, p. 33): 'life', 'bodily health', 'bodily integrity', 'senses, imagination, and thought', 'emotions', 'practical reason', 'affiliation', 'other species', 'play', and 'control over one's environment' (pp. 33–34).

These ideas make up an attractive ethical and political normative framework that could serve to evaluate new information and enhancement technologies, focusing in particular on the issues of well-being and *justice* they raise. We could ask if and how these technologies promote our human capabilities, and therefore our human *dignity*. For example, how do new social networks (re)shape our capability of affiliation? How does new information technology influence our relation to animals and to nature?

Moreover, if we combine the CA with what I have done in this book, it seems that *vulnerability transformations* can be evaluated by asking the question in terms of capabilities: do these transformations enable people to live a dignified and flourishing life, that is, do they enhance people's central capabilities? This seems to give us (1) a clear normative focus and (2) a more practical, operational way of understanding human vulnerability that could be used by policymakers. In the following pages, I explore what Nussbaum's version of the CA implies for the discussion about justice, enhancement, and vulnerability.

For this purpose, let me first reconnect to the previous section. Nussbaum takes care to stay close to the liberal tradition and develops her view in dialogue with the liberal-contractarian and in particular *Rawlsian* tradition. For example, she *claims* to refrain from offering a comprehensive, rich vision of the good life (Nussbaum 2011, p. 19) and from offering a particular doctrine that builds on 'any particular religious or metaphysical view' (p. 79). She also engages extensively with Rawls's version of social contract theory—especially in *Frontiers of Justice* (Nussbaum 2006). Here, I will not outline all the similarities and differences, but let me indicate some ways her theory differs from Rawls's.

Sufficitarian. In contrast to Rawls's prioritarianism (embodied in the difference principle), Nussbaum's list of capabilities is to be interpreted in a sufficitarian way. Applied to the issue of vulnerability, this means to ensure that no-one falls below a particular threshold of vulnerability, understood in terms of capabilities. With regard to enhancement, it does not mean to make sure that everyone has a decent 'genetic' minimum (as I argued, this involves all kinds of problems) but that social and technological arrangements are such that everyone is above a particular threshold of capability.

Procedural and not procedural. In Nussbaum's (and Sen's) capability approach (Nussbaum and Sen 1993; Nussbaum 2006), it seems that there is no procedure but rather a list of capabilities. However, Nussbaum argues that these capabilities could become part of an 'overlapping consensus' (Nussbaum 2011, p. 79). To the extent that the capabilities are formulated at a very general level and need political deliberation to gain substantial meaning, the approach *is* procedural.

Outcome-oriented. Capabilities can be regarded as a kind of outcomes, not resources. As I said, the capabilities approach was meant to include far more than resources. The Rawlsian theory of justice, by contrast, seems to be about resources: the principles of justice help us to achieve a fair distribution of resources.[6]

[6] The issue concerning resources versus goals reminds me of the debate between Dworkin and Sen on equality. Dworkin argued in *Sovereign Virtue* (2000) for equality of resources (a resource-based and procedural approach), whereas Sen (and later with Nussbaum; see Nussbaum and Sen 1993) is concerned with the goal of giving people capabilities (an outcome-based approach) (Sen 1992, 1995). For Dworkin, the issue is about resources, not outcomes—which is also the reason why Dworkin resists the term luck egalitarianism.

In sum, Nussbaum's version of the CA suggests an alternative, non-Rawlsian theory of justice to evaluate new information technologies and human enhancement, which still belongs to the liberal-philosophical tradition but is different from the Rawlsian theory and which has its own merits. First, with regard to the discussion about human enhancement, it avoids a direct discussion about what needs to be redistributed (the resources) since, in contrast to Rawls's theory, it is outcome-oriented: what matters are the capabilities or—in this discussion—the vulnerabilities understood in terms of (in)capabilities. Second, it includes a democratic-procedural aspect but in a less abstract way than in Rawls. We can agree with Nussbaum that capabilities need to be interpreted in particular contexts. If an 'overlapping consensus' is needed, then this means a consensus in a particular community or culture. Nussbaum stresses 'the equal worth and dignity of all persons' (Nussbaum 2011, p. 106), but at the same time, she wishes to respect the 'cultural or religious or ethnic or political identities' that constitute 'the spaces within which people express themselves in accordance with their choices' (p. 107). Nussbaum also argues that the capabilities list is 'open-ended and subject to ongoing revision and rethinking' and formulated in an abstract way to allow 'specifying and deliberating by citizens, legislatures, and courts' that takes into account histories and circumstances (p. 108). Third, the tension between freedom and justice is somewhat lessened, since the capability approach can formulate its normative concern in terms of justice and in terms of freedom.

For the discussion about new information technologies, this approach means that we focus less on the devices (the resources) and more on how they (may) change people's lives: their health, their social affiliations, and other capabilities. Moreover, it allows room for cultural differences with regard to this evaluation. Electronic devices may look the same everywhere on the planet, but ethical evaluation, interpretation, and the implementation of norms will always be partly dependent on the cultural and political context.

For the discussion about human enhancement, this implies that we should talk less about genes and more about what is done, or might be done, to our vulnerabilities and our capabilities. In addition, the focus on interpretation could be expanded

---

For Dworkin, the resources or 'initial endowments' that need to be equalised in order to allow people to face luck and uncertainty do not include what he calls 'talents'. The 'income-talented' would still have more income than others once the 'social' game is played (Dworkin 2002). Dworkin's resources seem not to include any 'natural endowments'. With regard to disability, he remarks that it is not the aim 'to make people equal in physical and mental constitution so far as this is possible' (Dworkin 2002, p. 123). He even suggests that we should treat 'genetic luck like other forms of luck' (Dworkin 2002, p. 125). (Dworkin actually means money and social position inherited from your parents, but I suppose he would make the same point about 'genetic potentials'.) While people should get equal resources that allow them to (hypothetically) insure them against luck, when the game plays out, the outcome is dependent on luck and talents. What matters for Dworkin is the initial starting situation, not the outcome. However, evaluating human enhancement in terms of justice seems to fall outside the scope of Dworkin's theory, since natural goods are excluded; it is a form of 'luck'. Now we could modify the theory to include natural-social goods in the initial resources, but then we encounter the problems of knowledge and freedom discussed above.

to a (more) hermeneutical approach to enhancement: it is up to interpretation if a particular enhancement is desirable, and it is also up to interpretation if a particular technology counts as 'enhancement' in the first place. Finally, we can also use the capability approach to argue for human enhancement in terms of freedom. One could try to justify human enhancement by arguing that freedom *demands* human enhancement: it can enhance capabilities of those with certain impairments but also of healthy persons and therefore make people more 'free'. This argument draws on a different definition of freedom: it is not about negative freedom, but about enabling people to exercise or expand their capabilities, and in that sense 'liberating' them.

However, problems still remain; let me mention some of them. First, since Nussbaum's version of the CA is sufficitarian, it misses an important part of the initial intuition about what may be wrong with human enhancement in terms of justice—indeed about what may be wrong with *any* wide gap between 'haves' and 'have-nots': the problem with 'human enhancement' visions is not only that some people might fall below a certain threshold of human dignity (their absolute position) but *also* that there is a wide gap between more enhanced and less enhanced (their relative position). The intuition is that even if a society were to make sure that everyone reaches a certain threshold, it is still *unfair* that there is such a wide gap. (Similarly, when it comes to justice as fairness, we want to care not only about 'absolute' vulnerability but also about 'relative' vulnerability.) As I suggested in the section on Rawls, prioritarianism offers a reason for caring about justice that concerns a problem 'internal' to the (enhancement) distribution, which is about relative position. This aspect of social and distributive justice seems to get lost if we use an 'external' criterion such as capabilities and set thresholds of capabilities. Perhaps a prioritarian principle can be integrated in Nussbaum's CA, but this requires more work. Moreover, when it comes to justice, we are sometimes confronted with the hard question of *whose* dignity,[7] vulnerability, and capability is more important and, in addition, with the difficult question which capabilities are more important—Nussbaum refuses to assign priorities to capabilities in the list.

Second, although Nussbaum allows for interpretation of capabilities, this might be understood as referring to interpretation by politicians and (other) citizens in an institutional context (e.g. in a legislative body), but politicians and citizens who are physically, socially, and psychologically far removed from the particular contexts, people, and experiences under consideration might give less attention to what capabilities mean to these people, to their concrete personal experience and existence. Nussbaum's emphasis on, and use of, narratives—for example, when she talks about women and the Internet, but also and especially in her earlier work on ancient Greek tragedy—shows how important personal experience is, but the precise relation between the CA as a political theory (about principles and laws) and existential vulnerability (about people and their lives) is still unclear. Can the law and the lawmakers take into account this more concrete, existential dimension? Can the law

[7] This problem is also raised by Nussbaum's arguments for inclusion of animals and future generations (Nussbaum 2006).

interpreter? What is the judge's role? What is the precise relation between the 'poetic justice' Nussbaum proposes in her earlier work and the justice in terms of capabilities she argues for in *Frontiers of Justice*? What is the difference between judges as interpreters and politicians as interpreters? What is the role of people outside these traditional political institutions with regard to the promotion of capabilities? What is the role of so-called civil society?

Third, in so far as Nussbaum's articulation of the CA is contractarian, it is problematic in the ways outlined in the previous pages: it presupposes that the social (and social justice) is something that needs to be justified and constructed. Therefore, it denies the reality of the social as given, in which forms of justice already grow—prior to (philosophical) rationalisations and justifications, and prior to evaluation and regulation. Finally, although work on technology and the CA is emerging, for example, on information technology and on the issue of 'ICT4D' ('ICT for development'), the relation between technology and capability is still under-theorised. In particular, the CA could benefit from more 'dialogue' with (nonmodern currents in) contemporary philosophy of technology. Let me explain this.

8.3.2.2 Beyond an Instrumental View of Technology: The Transformation of Vulnerabilities and Capabilities

The CA shares with all mainstream political and ethical theories an 'externalist' or 'instrumentalist' view of technology. Technology is seen as a means (a resource) to achieve aims (capabilities). With regard to enhancement and information technology in relation to human development, this means that according to the CA, it does not matter *what* people get but what they can effectively do with it. What matters are *human* dignity, *human* development, *human* freedom, and *human* empowerment. But this does not only raise questions about the relation between humans and other beings such as animals (a question Nussbaum discusses); it also raises the problem of *the relation between humans and technology* (a question that receives less attention). Nussbaum and others presuppose that there is no intrinsic, internal relation between humans and technology. We need technology for human development, but it is a resource. According to this view, humans use technology and do (good and bad) things to one another *by means of* technology, but what the human is and what human dignity requires is untouched by technology. The definition of the good is independent from technology; the end is independent of the means. Ethical and political principles can be formulated without taking into account the technological and material; they only come in when the 'empirical' world is attended to and when the *application* of the principles is at stake (to make this world a better place).

However, in the previous chapters, I have questioned this view of the relation between human being and technology. In tune with the transhumanist view that humans have always been 'enhanced' by technology and in line with currents in contemporary philosophy of technology that focus on how the human is changed by technology, I have argued (in Chaps. 2 and 4) that we must accept that *human being and human vulnerability have always been transformed by technology*.

With regard to the capability approach, this means that the definition of concept of human dignity itself should not rely on the idea of a fixed human nature. As Bostrom has argued, we can also talk about posthuman dignity, since 'what we are is not a function solely of our DNA but also of our technological and social context' (Bostrom 2005, p. 213). Moreover, what we are is always a matter of *interpretation*. Therefore, if human dignity is indeed a guiding ethical and political principle, then finding out what it means should include reflection on technology (next to and in relation to biology, society, and culture). It means, among other things, understanding how technology has already transformed our vulnerability (and indeed human being) and how it might transform our vulnerability in the future—'posthuman' or not.

Furthermore, a noninstrumental view also has implications for our understanding of capabilities. As I have argued in a recent chapter, understanding capabilities in relation to technology means that capabilities 'are not fixed but change together with our technological and social context', that they 'are *already* changing and have always changed' (Coeckelbergh 2011, p. 86). The very meaning of the capabilities changes as technologies reshape the human. Therefore, evaluating new technologies means evaluating the human *and* our ethical tools at the same time. The meaning of the capabilities such as 'bodily health', 'affiliation', and 'control over one's environment' changes when our technologies significantly transform our biological and social vulnerabilities.

For example, our use of social media and mobile electronic devices is changing what we mean by having social affiliations. Consider the case of a person who does not have 'real' face-to-face contact with people for a long time but is still connected to people via the Internet and other new media. A 100 years ago, physical isolation was also social isolation. Physical isolation would raise an insurmountable problem with regard to a person's capability of social affiliation (it was a punishment or a catastrophe). Even if such a person had enough material resources to sustain physical life, her situation would be without dignity since she would live without others. It would be unjust to do this to anyone; even prisoners are rarely isolated. It is a severe punishment; it is a kind of torture. Today, we also consider the situation undignified and unjust, perhaps, but if the person has access to our contemporary communication technologies, we may interpret her predicament differently and take into consideration that the person still has 'online' contact with others. Even if we consider such a life worse than another (in which case we have to argue why we think this is so), we are still challenged not so much to apply our moral principles but rather to think about what they mean with regard to these new lives, transformed as they are by the new technologies and new media. For instance, we must ask what social affiliation (and indeed living a dignified human life) means in the age of information technology and social media. How do they transform our social vulnerability and social capability? This means asking precise questions about particular practices, particular technologies, and about the meaning of what we believe is valuable. What is 'social affiliation'? Can friendship be sustained online? Is face-to-face contact better? Is it necessary? Should we use a webcam? Ethical and political thinking about new technologies and about human enhancement cannot do without such an (already normative) hermeneutics of techno-human change, understood as a hermeneutics of vulnerability or capability transformations.

8.3.3 Ecologies of Vulnerability and Ecological Justice

So far, my discussion of justice and fairness in relation to human enhancement and vulnerability has not left the boundaries of the so-called human world. But this limitation is highly problematic. For example, does it mean that we can say *nothing* about justice with regard to our relation to animals and the environment? Consider animals. Can we say nothing about *their* vulnerability and *the possibility of their enhancement*? Is our vulnerability independent from their vulnerability? Is our vulnerability entirely *different* from their vulnerability? Is our relation to them not a matter of justice? Is this area situated *outside* the sphere of justice?

In this section, I will construct an argument for a more ecological view of justice and vulnerability by responding to contractarianism.

At first sight, contractarianism (including Rawls's version of it) seems unable to cope with ethical issues concerning the non-human in a way that corresponds to our contemporary intuitions. This is also Nussbaum's suggestion in *Frontiers of Justice* (2006). Animals, for instance, are excluded from the social contract. (In fact, all beings that are not fully rational beings seem to be excluded from it, including some humans.) And in Rawls's case, animals are not one of the parties in the original position, who decide upon the principles of justice.

Confronted with this difficulty, some may decide to immediately drop contractarian theory of justice, to study Nussbaum's modification of contractarianism, or to adopt more ecological ways of thinking suggested by influential currents in environmental philosophy. I sympathise with the third route, but I first wish to indicate a possibility that arises within the contractarian framework—if it is broadly interpreted and significantly modified. There is a way to open up the contractarian theory of justice to non-humans using an important contractarian intuition and assumption: the problem of distributive justice only arises within a scheme of social cooperation. Social cooperation entails dependency: the cooperators are dependent on one another. Our shared vulnerability and dependency is the basis of our cooperation. But if this is true, then we do not have a good reason to restrict this sphere of dependency and cooperation to 'society', the 'merely' human world. Humans crucially depend on non-humans for their existence: they need animals, plants, 'raw materials', air, and the ecosystem. And given the enormous power that humans have to change their environment and therefore to change the lives of other beings, this dependency is a mutual one. Other beings are also dependent on us. The result is the growth of forms of (quasi-)social cooperation between humans and non-humans. Many relations between humans and non-humans can be interpreted as 'cooperative' in the sense that we do things together in order to cope with our vulnerability: our 'cooperation' arises from our co-vulnerability. If this makes sense, then according to this broad interpretation of 'social cooperation' (see also Coeckelbergh 2009), the scope of a theory of justice should be widened too. To the extent that coexistence means 'cooperation', problems of justice and fairness can arise between all coexisting life forms, objects, and ecosystems. (Of course, this means we must drop the rationality requirement, at least as a requirement for all the 'parties' to the social contract. But when it comes to the procedure, it is still us—rational humans—who think about ecological justice.) Our non-human 'cooperators' also fall within the scope of justice.

Now if we take seriously the concept of existential vulnerability, this argument can be further elaborated. Recall that in Chap. 6 I wrote that there are 'ecologies of vulnerability'. This is an implication of the relational nature of existential vulnerability. Possibilities of violence and possibilities of good only arise in relation to other humans, objects, and systems. Human being as relational existence has all these different aspects (it is social, technological, etc.) and must be understood not only as dynamic but also as holistic. There is mutual and plural dependency, which renders us vulnerable in a mutual and plural way. We share 'biological' vulnerability, but 'biology' should be understood in an environmental way. Shared 'biological' vulnerability is not so much about 'having' similar 'bodies'; rather, it is about relating to one's environment in similar ways. The concept of vulnerability enables us to go beyond the Cartesian presuppositions of an animal ethics and an animal politics based on having similar 'bodies' and 'minds'. Sharing vulnerabilities is a non-dualistic notion; it is about relations. It is about sharing dependencies. It is about being part of the same ecology. If something changes in such an ecology of dependency and vulnerability, other things will change as well. If something changes in the vulnerability of an animal, then this will change the vulnerability of animals and humans related to it. If something changes in human vulnerability, then this will have consequences for the vulnerability of other beings and things. Existential human vulnerability is not the only kind of vulnerability there is and, by its very definition, does not stand alone but is related to these other, non-human vulnerabilities. As said, human vulnerability is always co-vulnerability.

To the extent that these ecologies of vulnerability can be interpreted as cooperative schemes, the question concerning (distributive) justice and fairness can and must be asked from a contractarian point of view. But of course, we can also take a noncontractarian view and drop the requirement of cooperation: then, coexistence and hence co-vulnerability is sufficient for broadening the scope of justice. This is what we may call a 'genuinely' ecological view of vulnerability and of human existence, since it moves away from the presupposition that 'first' there are entirely separate entities that 'then' cooperate. Instead, it understands both our vulnerability and the way we cope with that vulnerability as intrinsically related to the vulnerability and anti-vulnerability strategies of other beings and things. It is also a genuinely *environmental* view of human being and of vulnerability: what we are and the kind of vulnerability we have is entirely dependent on our environment; there is no 'non-environmental' core that remains untouched by the environment. Finally, it means that questions of justice in relation to human vulnerability must be asked in a way that recognises the relational, ecological, and environmental nature of that vulnerability. It means that we cannot ask 'Is this vulnerability transformation fair' without considering the fairness of how other, non-human vulnerabilities are transformed by it.

With regard to enhancement technologies, this question of 'ecological' justice can be asked at a general meta-level: Is it fair that human vulnerability transformations (e.g. technological ones) reduce human vulnerability but at the same time increase the vulnerability of other beings and ecosystems? But if what I said about cultures and transformations of vulnerability is correct, we should ask more particular questions: Is it fair that *this* particular vulnerability transformation, for example, enhancement of crops by genetic manipulation, reduces human food vulnerability (for *these* particular people) and transforms the vulnerability to disease and pesticides in a particular way (for *these* plants), but also transforms the vulnerability of that particular ecosystem, making particular (other) plants more or less vulnerable to a particular chemical, and perhaps increasing the vulnerability of *other* people? Is it fair that we transform an animal's vulnerability in a particular way (by genetic means) in order to make ourselves (humans) less vulnerable to financial loss or to loss of a particular way of eating and a particular way of life? To ask such questions about justice with regard to the relation between humans and non-humans in this more 'ecological' manner is unusual, and to ask them in terms of vulnerability may strike us as somewhat exotic. But it makes sense to ask these questions—indeed we *must* ask them—if we take seriously the novel conceptual framework to descriptive and normative environmental politics and ethics suggested in this book.

Moreover, these questions also make us think about what kind of beings we want to be (normative anthropology). If we consider being fair as an element that should be part of humanity's moral identity, then we should not limit our efforts to promote justice to 'the human world'—if that means anything at all. And if my analysis of existential vulnerability as co-vulnerability and ecological vulnerability is correct, the 'mere' human world does not even exist. As far as vulnerability is concerned, human being cannot understood apart from non-human being. Therefore, normative anthropology of vulnerability is always also normative ecology of vulnerability. A reflection on the ethics and politics of human being (normative *anthropo-logos*) also implies reflection on the politics and ethics of the 'household' and the environment in which we live, on which we depend, and which we continue to transform (*eco-nomos* and *eco-logos*)—as much as it transforms us. When we think and say something about how the human should be, at the same time, we think and say something about how the world should be, including the non-human. In this sense, any descriptive and normative philosophical anthropology must be at the same time an *environmental* philosophy. (Note that, like environmental philosophy, this also means that our normative anthropology of vulnerability has an aesthetic aspect. For example, it also means that we think about how the world should *look* like. Apart from politics and ethics, our normative discussion also includes the aesthetics, the *beauty* and *style* of human being and world, which I shall discuss in the next chapter.)

8.4 Transformations of the Political in the Information Age: Public and Private

To end this chapter, let me discuss the relation between the political, vulnerability, and technology by commenting on Hannah Arendt's view of politics, in particular her distinction between the public and the private realm. This will also help me to further elaborate the arguments I made about the relation between freedom and vulnerability and about the relation between transcendence and vulnerability.

8.4.1 Standard View: The Public Sphere as a Sphere of Invulnerability

Our thinking about politics and human being is (partly) rooted in ancient Greek thinking and practice. In so far as they are still Platonic, our anthropologies and spiritualities are vulnerability-phobic and seek transcendence, and they are continued and cultivated in modernity. As I suggested earlier, modern technologies, and especially enhancement technologies as *transhumanist* technologies, can be considered as anti-vulnerability tools we use to try to release ourselves from dependency and vulnerability, especially from 'that vulnerable thing' par excellence: the human body. And I have also shown that the search for invulnerability is already present in the monotheistic religions—partly due to Platonic influence, partly because monotheistic religion was always and intrinsically dualistic and earth-unfriendly, dividing up the world between creator and creation, and seeking escape from the human condition. We do not want to return to the dust—and in saying this, we consider ourselves to be separate from the earth.

But what is the relation between technology, vulnerability, and *politics*? When we try to think about the political, we meet the Greeks again. In *The Human Condition* (1958), Hannah Arendt has articulated a view of ancient (Greek) and modern politics that may help us to expose and discuss the assumptions we make about vulnerability, politics, and democracy.

Following the ancient Greeks, in particular Aristotle, Arendt makes a distinction between a public realm and a private realm. The centre of the private realm is the home (*oikos*) and the family, whereas the public realm allowed one to have a 'second life' as a *bios politicos* (Arendt 1958, p. 24). The first domain is related to the 'maintenance of life', whereas the second relates to 'a common world' (p. 28). The first is the sphere of necessity (the household was born of necessity); the latter is the sphere of freedom. The first was the condition for the latter: 'the mastering of the necessities of life in the household was the condition of freedom of the polis' (pp. 30–31). Freedom depended upon wealth and health: 'To be poor or to be in ill health meant to be subject to physical necessity, and to be a slave meant to be subject, in addition, to man-made violence' (Arendt 1958, p. 31).

Thus, inspired by Greek thinking, Arendt connects political action with freedom from necessity and with mastery. In this way, she constructs the political and public sphere as what we may regard as a 'sphere of invulnerability', in which we are free from bodily needs and from violence. In order to become political, we leave behind the *bios*. The ideal of freedom and politics she promotes is an ideal of *transcendence*:

> To be free meant both not to be subject to the necessity of life or to the command of another *and* not to be in command oneself. It meant neither to rule nor to be ruled. Thus within the realm of the household, freedom did not exist, for the household head, its ruler, was considered to be free only in so far as he had the power to leave the household and enter the political realm, where all were equals. (Arendt 1958, p. 32)

Arendt then argues that in the modern age, this distinction is crossed and blurred. A realm emerges (the 'social realm') which is neither private nor public. In the modern world, private matters become a collective concern. Everything becomes part of 'the life process' (p. 33). In this way a 'curiously hybrid realm' arises where 'private interests assume public significance' (p. 35). Whereas before the good life was about mastering life's necessities, now we are 'bound to the biological life process' (p. 37). Arendt seems to regret this impurity. We can interpret this response not as a denial of vulnerability, but as a denial that *politics* should involve vulnerability. The pollution of the public sphere by private concerns, by the household, is viewed as bad because it renders us vulnerable again. We can no longer escape from necessity or escape from our bodily needs. They are a 'burden' (p. 37) and make us vulnerable to the violence of others.

The relation Arendt draws between politics and transcendence becomes clear when she touches upon the issue of immortality. Politics, to appear in public, is a means to leave the earth and seek a more permanent state:

> Through many ages before us—but now not any more—men entered the public realm because they wanted something of their own or something they had in common with others to be more permanent than their earthly lives. (Thus, the curse of slavery consisted not only in being deprived of freedom and of visibility, but also in the fear of these obscure people themselves "that from being obscure they should pass away leaving no trace that they have existed.") There is perhaps no clearer testimony to the loss of the public realm in the modern age than the almost complete loss of authentic concern with immortality. (Arendt 1958, p. 55)

For Arendt, it seems, politics is an anti-vulnerability strategy. By entering the public realm, we give ourselves 'the possibility of immortalising':

> For the *polis* was for the Greeks, as the *res publica* was for the Romans, first of all their guarantee against the futility of individual life, the space protected against this futility and reserved for the relative permanence, if not immortality, of mortals. (Arendt 1958, p. 56)

Like transhumanists who hope that technology will allow us to live forever, these 'men' were seeking 'earthly immortality' (p. 56). They wanted to gain freedom from labour and from consumption, from the metabolism of life, and indeed from the body. The body was hidden: 'it has always been the bodily part of human existence that needed to be hidden in privacy' (p. 72). Women and children were also hidden: they belonged to the private sphere, where life was 'devoted to bodily functions' (p. 72). But today, 'the necessary, the futile, and the shameful' no longer have their 'proper place in the private realm' (p. 73). Thus, we can interpret politics as a strategy not so much to deny our vulnerable existential condition or to openly fight it, but to *hide* it. In ancient times, our struggle with vulnerability *could not be openly* done; it had to be hidden from public perception. In the contemporary world, by contrast, vulnerability becomes publicly visible. When the distinction between the private and the public blurs, vulnerability shows.

For democracy, this means that rule by the people (or today by those who call themselves representatives of the people) also becomes a matter of *management*: household, economy. Politics in modern nation states becomes political economy, that is, the rule of the national household. Whereas before 'private' matters could not appear in public (e.g. they would be matters dealt with in companies, literally 'those who eat the same bread'), now economic and financial matters are the core 'business' of politicians. Politicians manage the state and—like companies and their advertisement departments—in a consumer society, they also try to manage the desires of the people. Political parties and politicians try to create a good *brand*. They aim to seduce people. And (populist) politicians try to represent in the sense that they try to collect the individual preferences of people: they are the visible hand next to the invisible hand of the market. Politicians become business people: they are busy with the household, with the economy of goods, money, and desire—with necessity rather than freedom.

Thus, if Arendt is right, in modern times, the word has been replaced by the flesh: we are now bound to the life process, to the metabolism. The sphere of politics has been replaced by the sphere of management, which is inhospitable to the word. Logos and reason is gone. The darkness has taken over the light. There is no hope of salvation from the earthly, bodily sphere. Women are shown everywhere, they can 'even' become 'politicians'. All workers are slaves; all consumers are children. The private and the public are mixed.

Contemporary information and communication technologies seem to promote this collapse of the public-private distinction, since they render the private public. Social networks are *privatisers*. The intimate is shown and thereby destroyed. Moreover, as I already mentioned when referring to Dreyfus's view, it seems that the Internet encourages the flourishing of *private* opinion as opposed to *public* reason. Cyberspace becomes one big household and one big family. E-democracy is not a public matter but a matter of (ac)counting and accumulation. Internet may make private property of individuals less important (more important is *access* to the network and the cloud, not property), but our private identities and our private lives become the property of large global companies. Our minds and bodies become information containers and information processors monitored by e-health institutions and by businesses. As I suggested before, we are the new slaves: our labour consists of producing data they can sell. We are milked for information. We become part of streams of information; we become part of a global *metabolism* of information. In other words, this is the party of necessity, not the feast of freedom. For politics à la Arendt, these are dark times indeed.

8.4.2 Alternative: Bio-politics and Info-politics. Bloody Democracy and Silicon Voices

Arendt rightly warns us of the possibility that everything becomes private if that means that nothing is common; we need a common world and we need to attend to public matters. Recent information technology that aims at 'personalisation' seems particularly dangerous in this respect, since it tends to create and promote not a *common* world but a *collection* of private 'worlds': we only get to see the information the search engine has selected for us on the basis of our *preferences*, we are occupied with our own 'friends', and we are living 'iLives'. Furthermore, it seems true that politics is turned into household management. Since the birth of the nation state, for example, it has been dealt with as a nation *estate*, a property and a territory that has to be managed. We are also lured into thinking of our nation as a company, which has to compete with other nation's companies on the global market. I sympathise with what could be further developed into an Arendtian critique of consumer society and the omnipresence of the private and the intimate.

However, I doubt (1) if the 'ancient' politics Arendt projects *ever* existed and ever *can* exist and (2) if the ideal of politics in terms of a *transcendent* public sphere Arendt defends is indeed something we should want. It seems more plausible that politics has always been 'hybrid': public and private, earthly and less earthly at the same time. The speech of politicians is always *about* something, in particular about human affairs. How unearthly can politics really get, if it is—as Arendt would concur—always connected to the common world? The common world is inevitable (also) material and technological. Moreover, perhaps we should prevent the disappearance of 'the public sphere', but how separate is this sphere? Surely it will always have to include the private to some extent. And can it disappear at all? 'Reality' as the common world will always break open again our private cocoons. In other words, Arendt's distinction itself must be questioned. Necessity, embodied existence, and private desire have always been part of the political. Arendt's effort to purify the political is futile: although we might want to question the current dominance of the private (in public) and of economical thinking (in politics) and take *distance* in this sense, we will never attain the degree of invulnerability from necessity she imagines. There will always be a 'social' sphere that is impure. Politics is also always bio-politics and info-politics.

Democracy means "giving the law to yourself" (autonomy), but it also involves "giving the bread to yourself" (*autopany*?). We need rules and we need bread. As Aristotle explained, the polis grew out of necessity. The flower of the polis (the word, the logos) is rooted in the soil of the flesh. The ideal of politics that emerges from Arendt's writings is too bloodless. If democracy is an ideal at all, then it has to be a 'bloody democracy', which does not hide existential vulnerability but lets it appear in public. If we listen to the voices of the oppressed, this should not be interpreted as listening to *reasons* alone but also as listening to real people and their sufferings. The word presupposes a voice, and the voice presupposes a throat and a body. If we forget this, we make laws that violate those bodies. If we forget the blood or try to expel it, the blood will come to us and flood the palaces of fake transcendence.

Moreover, the material, which is usually connected to necessity, is also part of the political. Arendt makes a sharp distinction between the immaterial sphere of freedom and politics and the material sphere of necessity. Admittedly, she sees the latter as a condition of the former. But her view that we first have to attend to the material side of life before we can enter the political is misguided. We cannot neatly separate the two spheres. As Latour argued (2004), we have to bring things into the collective, into democracy. Arendt's view seems to support the post-industrialist idea that science and technology allow us to have things that take care of our necessity (electronic gadgets, robots, etc.) and to live comfortable lives, and that this gives us the 'free time' to do what we really want. But in practice, this strict separation between necessity and meaning cannot be sustained. Things always play a role in our lives and our engagement with them contributes to their meaning. In this sense, we have to listen to silicon voices too. If we wish to build and (re)shape the common world, we have to attend to its biological and material aspects. Living beings and things are not resources; they are part of our body. Animals and technologies are part of the *body politic*. What happens to them happens to us. And with regard to vulnerability: what *might* happen to them might happen to us.

In sum, the ideal of politics and freedom promoted by Arendt's analysis tries to hide vulnerability, but it will always show up again and the 'earthly immortality' we might seek in the political sphere will always depend on bodies and things. The 'social sphere' is perennial—at least as long as we remain humans: humans who are social by nature, and who suffer but also appear and flourish *as* technological and biological beings. As persons, we should stop trying to kill the zombies and face our own mortality. As politicians, we should start *governing* the household rather than believing in invisible hands, thereby taking into account the vulnerability of the people—that is, the vulnerability of *all* people and the vulnerability of the beings, tools, and environments related to them. We should not cover up vulnerability, but evaluate and discuss it within the framework of a democratic politics of vulnerability—keeping in mind, of course, that democracy itself is a highly vulnerable and unstable political institution, and that the *demos* and its vulnerability must be understood in ecological terms.

References

Arendt, Hannah. 1958. *The human condition*. Chicago: The University of Chicago Press.

Beck, Ulrich. 1992. *Risk society: Towards a new modernity*. Trans. M. Ritter. London: Sage.

Bostrom, Nick. 2005. In defence of posthuman dignity. *Bioethics* 19(3): 202–214.

Coeckelbergh, Mark. 2009. Distributive justice and cooperation in a world of humans and non-humans: A contractarian argument for drawing non-humans into the sphere of justice. *Res Publica* 15(1): 67–84.

Coeckelbergh, Mark. 2011. Human development or human enhancement? A methodological reflection on capabilities and the evaluation of information technologies. *Ethics and Information Technology* 13(2): 81–92.

Coeckelbergh, Mark. 2012. *Growing moral relations: Critique of moral status ascription*. Basingstoke/New York: Palgrave Macmillan.

Diamond, Larry. 2010. Liberation technology. *Journal of Democracy* 21(3): 69–83.

Donaldson, Sue, and Will Kymlicka. 2011. *Zoopolis: A political theory of animal rights*. Oxford/New York: Oxford University Press.

Dworkin, Ronald. 2000. *Sovereign virtue*. Cambridge, MA: Harvard University Press.

Dworkin, Ronald. 2002. Sovereign virtue revisited. *Ethics* 113: 106–143.

Farrelly, Colin. 2004. The genetic difference principle. *The American Journal of Bioethics* 4(2): W21–W28.

Farrelly, Colin. 2005. Justice in the genetically transformed society. *Kennedy Institute of Ethics Journal* 15(1): 91–99.

Foucault, Michel. 1977. *Discipline and punish: The birth of the prison*. Harmondsworth: Penguin.

Foucault, Michel. 1988. Technologies of the self. In *Technologies of the self: A seminar with Michel Foucault*, ed. L.H. Martin, H. Gutman, and P.H. Hutton. Amherst: The University of Massachusetts Press.

Hayek, Friedrich A. 1976. *Law, legislation and liberty: A new statement of the liberal principles of justice and political economy, vol. 2. The mirage of social justice*. London: Routledge & Kegan Paul.

Latour, Bruno. 1993. *We have never been modern.* Trans. C. Porter. Cambridge, MA: Harvard University Press.

Latour, Bruno. 2004. *Politics of nature: How to bring the sciences into democracy.* Trans. C. Porter. Cambridge, MA: Harvard University Press.

Leiter, Brian. 2010. Cleaning cyber-cesspools: Google and free speech. In *The offensive internet: Privacy, speech and reputation*, ed. S. Levmore and M.C. Nussbaum. Cambridge, MA: Harvard University Press.

Levmore, Saul, and Martha C. Nussbaum (eds.). 2010. *The offensive internet: Privacy, speech and reputation*. Cambridge, MA: Harvard University Press.

Lewontin, Richard. 2001. *The triple helix: Gene, organism, and environment*. Cambridge, MA/London: Harvard University Press.

Lindsay, Ronald A. 2005. Enhancements and justice: Problems in determining the requirements of justice in a genetically transformed society. *Kennedy Institute of Ethics Journal* 15(1): 3–38.

Moss, Lenny. 2003. *What genes can't do*. Cambridge, MA: MIT Press.

Mouffe, Chantal. 2000. *Deliberative democracy or agonistic pluralism*, Political science series 72. Vienna: Institute of Advanced Studies.

Nussbaum, Martha. 2006. *Frontiers of justice: Disability, nationality, species membership*. London/Cambridge, MA: Harvard University Press.

Nussbaum, Martha. 2011. *Creating capabilities: The human development approach*. Cambridge, MA/London: The Belknap Press of Harvard University Press.

Nussbaum, Martha, and Amartya Sen. 1993. *The quality of life*. Oxford: Clarendon.

Rawls, John. 1971. *A theory of justice*. Cambridge, MA: Harvard University Press.

Salvi, Maurizio. 2002. Genetics' dreams in the post genomics era. *Medicine, Health Care and Philosophy* 5: 73–77.

Sen, Amartya. 1992. *Inequality re-examined*. Cambridge, MA: Harvard University Press.

Sen, Amartya. 1995. Equality of what? In *Equal freedom*, ed. S. Darwall, 307–330. Ann Arbor: University of Michigan Press.

Sen, Amartya. 1999. *Development as freedom*. Oxford/New York: Oxford University Press.

Chapter 9
Normative Aesthetics of Vulnerability: The Art of Coping with Vulnerability

9.1 Introduction

Usually ethical and political-philosophical discussions do not involve a discussion of aesthetics. The latter is understood as being concerned with matters of beauty and taste, whereas ethics is seen as being about right and wrong or about good and virtue. Yet in this chapter, I will argue that when it comes to evaluating vulnerability transformations and indeed when it comes to thinking about what kinds of humans we want to be, such a sharp distinction between ethics and aesthetics is unfruitful for two main reasons. First, the aesthetics of vulnerability is *normative* in various ways and therefore deserves its place within a normative anthropology of vulnerability broadly conceived. Second, ethics itself, and therefore also the ethics of vulnerability, can be understood as a kind of art. Thus, it is not only the case that normative anthropology is 'also' about beauty; I will argue that it is also and crucially about coping with vulnerability as an art, understood in the sense of a craft or *technè*, which requires skills. There is not only an extrinsic but also an intrinsic connection between ethics and aesthetics.

First, I will explore extrinsic connections, then I will discuss what I take to be an intrinsic connection between the ethics and aesthetics of vulnerability. In addition, I will explore some 'aesthetics'-related themes that are relevant to the present discussion about vulnerability: fate and tragedy, architecture and design, and the appearance of animals and robots. Then I will recommend that we seek an ethics-aesthetics that moves 'beyond the human': not by trying to design posthumans but rather by taking seriously the environmental, ecological nature of vulnerability.

M. Coeckelbergh, *Human Being @ Risk: Enhancement, Technology, and the Evaluation of Vulnerability Transformations*, Philosophy of Engineering and Technology 12,
DOI 10.1007/978-94-007-6025-7_9, © Springer Science+Business Media Dordrecht 2013

9.2 The Idea of a Normative Aesthetics of Vulnerability: From Representations of Vulnerability to the Art of Coping with Vulnerability

There are several ways to conceive of the relation between aesthetics and existential vulnerability. Some of them are directly relevant to the ethics and politics of new technologies and of human enhancement; others are not 'ethical' or 'political' in a straightforward way but nevertheless invite questions that help us to further develop the project of a normative anthropology of existential vulnerability.

First, we may think of works of art that explicitly or implicitly explore the origin, experience, or limits of human vulnerability. Perhaps most good art does so implicitly in narratives of human lives and images of human being. For instance, with regard to human enhancement, we may consider the work of artists who explicitly explore our relation to contemporary or new technology, such as Stelarc[1] or Hans Op de Beeck,[2] who express and explore experiences of technological vulnerability transformations by using the language of 'extension'. We could also consider the work of artists who explicitly reflect on what they consider to be the 'posthuman future' and call themselves transhumanist artists, such as Natasha Vita-More.

Second, we may understand some works of art as explicit elaborations of, and 'reflections' on, more common and widely shared aesthetic experiences of vulnerability—including experiences related to contemporary or new technology. Artists may be more sensitive to issues related to vulnerability transformations than most of us, and their work may help us to look at our vulnerability experiences and our relation to technology from a distance. But if and in so far artists play this role, then what they do is more the work of a midwife than the work of the craftsman. What we may call an 'aesthetics of vulnerability' is already at work in our lives and in the way we talk about things: we can and do describe and reflect on vulnerability in terms such as beautiful and ugly, the sublime, good taste, and bad taste since we partly *experience* it in this way. The aesthetical quality of these experiences can be made more explicit by using well-known concepts from (philosophical) aesthetics. For example, our dependence on technology, our being-vulnerable-to-technology in particular situations, may be described in ways that invite the use of the aesthetic term 'sublime'. We may also ascribe beauty to the one who is threatened and ugliness to the threat. And we experience our vulnerability differently depending on how we perceive aesthetic features of our environment since this perception changes our relation to the world and hence our vulnerability experience. Some features appear as more threatening than others. We also draw on images of the past. For instance, in the post-9/11 period, many US citizens experienced their vulnerability partly in relation to images of the collapse of the WTC towers. More generally, the aesthetics of vulnerability and risk partly depends on the aesthetics of threat, attack, and disaster, and—more generally—on an aesthetics of *violence*. Our relation to the

[1] See http://www.stelarc.va.com.au/

[2] See http://www.hansopdebeeck.com/

world is mediated by fictional and nonfictional violence shown and described by various mass media. Vulnerability is partly constituted by how you perceive the world, which depends on your culture, society, and person. (Consider again my previous remarks on how the concept of vulnerability relates subject and object.) How is pain experienced in a Christian culture? How is the possibility of death experienced by people who do not believe in an afterlife?

For the aesthetics of vulnerability, *appearance* matters. There is no 'objective' reality, or if there is one, we do not have access to it. The features of our environment that appear to us as threatening are not 'objective' features but are mediated by our personal and cultural horizon. The scientific, 'objective' way of looking at things is the result of one kind of mediation (by modern science and education). Vulnerability and risk, understood as the possibility of violence, may arise from the perception of particular aesthetic features—which depends on culture and individual history. When in the aftermath of a nuclear disaster people look at a nuclear power plant, they do not see 'objective information' about nuclear risk. They do not see 'facts' about nuclear risk. They do not see 'technological installations' for the generation of electricity. They see *risk*. They feel a threat. The cooling towers—which 'objectively' have no intrinsic connection to the nuclear at all—become the towers of doom. They come to represent nuclear risk; they 'become' the risk. Similarly, in a particular situation, a particular face and figure in the street may be sufficient to evoke the possibility of violence—even if the person has very different intentions. A *shadow* of the figure is enough. We do not see a person or a human being. We see a *threat*. (To see a 'human being' in such a situation would be a very particular, so-called 'objective' way of looking at the threatening other, which requires not only a lot of training but also a kind of Gestalt switch—in these cases, we have to exert effort *not* to fear.) From 'the outside', we see a victim and a criminal. Risk and vulnerability are embodied: they are in the posture, the gesture, the facial expression, the way of walking, and the sounds of the body (e.g. the heavy, deafening heartbeat; the nervous footsteps; the startled breath). Levinas based an entire approach to ethics on such a vulnerability experience: vulnerability *shows* itself in the face of the other; it *speaks* to us. We *sense* risk and vulnerability.

Indeed, we should add other senses too and not restrict the phenomenology of risk and vulnerability to vision. What is the *sound* of the threat? What is the *smell* of vulnerability? How does risk *touch* me? Consider the sound of thunder (the possibility of being hit by lightning), the sound of a mosquito (the possibility of disease), the clicking sound of a Geiger counter (the possibility of ionising radiation), the sound of a warning message from the computer (the possibility of a computer virus and a computer crash), the smell of decaying food (the possibility of disease), the feeling of heat of a fire on your skin and on your eyes and the smell of smoke (the possibility of burning and suffocation), the experience of being 'touched' by a heart attack (the possibility of sudden death), the smell of a corpse or carcass (the possibility of *your own death*), etc. In our culture, we often try to make risks *visible* since we tend to rely less on our other senses, but this does not mean that we have entirely lost other 'risk senses'.

Third, there is also a *normative* aesthetics of vulnerability. The experiences summed up here are neither culturally nor normatively neutral. They already include an evaluation, an attitude, and a way of coping with the risk. How *should* we view, listen to, smell, and taste vulnerability?[3] Is it *wrong* to associate particular looks with risk, and what would be *right*? Should we try to make death invisible without smell? Should we not touch it? What is the taste of immortality? Can the smell of death be sweet? How does invulnerability taste? Is its eternity a silent sky or a roaring river? All these different descriptions and experiential possibilities include evaluations and can be part of a more comprehensive normative-anthropological project.

With regard to the evaluation of new technologies, we may ask if and how the new technology redefines our (categories of) taste, how it redefines the beautiful, and how this impacts on our vulnerability. For example, does new information technology really turn us into *screenagers*, and does this mean that vision becomes even more important in our culture at the expense of other sense-abilities and sensibilities? What sounds, smells, and touches are covered by the screen? What does the screen show, and what does it hide? What does the dominance of the visual mean for vulnerability? Do we become globally and electronically extended *eyes* or *retinas* (rather than extended skins, eardrums, or tongues?) Do we want *something else* than screens as information and communication media?[4] Here is an another example: if cosmetic surgery technologies and mass media images of 'perfect' people and stories of supermodels are widely available, beauty is redefined in a particular way and transforms the vulnerability of some people, in particular those who were already vulnerable when it came to their (view of their) looks. And with regard to transhumanist human enhancement, one question is, if and to the extent that the design of posthumans is possible and desirable at all, how should posthumans look, feel, touch, hear, and smell? Perhaps they would have *other* senses too (with the help of technology). To the extent that our technologies enable more control over our physical appearance and over our senses, the domain of the normative is also extended to these realms. In so far as it is possible to 'design' our senses, what kind of ears, eyes, skins, and tongues do we want? What kind of *extra* sense do we want, if any? What does this tell about what kind of human we want to be? What does this tell about how we want to relate to our environment? Finally, artists such as Stelarc or Koert van Mensvoort do not only 'describe' but also partly project and explore visions of how we *should* deal with technology and how the human *should* be, including how we should look, how technology should look, and how we-with-technology (or we-as-technological) should look, feel, etc. Their art is always normative, even if they may emit ambiguous and vague normative messages (and it is not sure if, in so far as they are artists, they *can* or should be clearer: too much clarity—like too much vagueness—may come at the expense of the loss of meaning;

---

[3] I refer to several senses to counter the contemporary obsession with sight and hearing alone. Levinas is very contemporary in that sense.

[4] For example, Koert van Mensvoort has suggested that we could use other things (in our surroundings) as media, such as fountains, plants, or other elements of our environment. See http://www.koert.com/work/

it may unnecessarily limit the space of the ethical-aesthetical imagination that art can help to open up). More generally, we may ask: what is a *beautiful* relation with technology, for example, enhancement technology? Do new technologies allow us to achieve beauty? Could enhancement technologies change our sense of beauty (not only our *view* of beauty) and, in fact, all our senses? What does this mean for *making sense* of the world? To change our sense of the world is partly to create a new world. To change our sense of ourselves is partly to create a new human being. The 'subjective' and the 'objective' merge.

Fourth, unless we, as a society and as a transnational community of citizens, participate in political decisions about human enhancement technology, others (e.g. professional politicians, civil servants, engineers, scientists, and business people) will decide the aesthetics of the future human being—'posthuman' or not. Surely, participation in the politics of aesthetics does not necessarily generate more beauty, let alone the most beautiful results. But public participation in ethical and aesthetical decisions about science and technology will at least guarantee that we will have some say on the ethical and aesthetical quality of our future world and being.

More, this discussion also suggests that ethics and aesthetics might be more closely related than they are often assumed to be. Given the normative and pervasive aspect of the aesthetics of technology, the beauty of our world and being should not be regarded as 'a mere matter of taste' if this means 'a matter of individual judgment'. Rather, we can and need to publicly discuss it as a social, communal matter. It is a matter of 'common' taste understood as a *sensus communis*. To arrive at a common sense, Kant recommended in the *Critique of Judgement* that we compare our judgment with that of others by putting ourselves in the position of everyone else. Whether or not this is possible or desirable, we can argue and discuss about our judgments. If we take seriously the question regarding the beauty and the sense of technologically transformed vulnerability, a normative aesthetics of the human and the posthuman understood as a normative aesthetics of vulnerability is not a matter of 'anything goes' but an urgent normative philosophical, ethical, and political issue. At stake are the beauty of the world we (will) have to inhabit and indeed the shape of the human being—a beauty and a shape which we do not fully control but at the same time *already* influence, evaluate, and live. Of course, this view generates the same problem as an ethics of future technology: what if our *aesthetic* values change in the future? And since taste tends to be more capricious than morals (change is faster, but that doesn't mean that it does not *matter* how we look), how can we ever know the look, sound, feel/touch, taste, and smell of the future human? For example, if we tried to imagine 'posthuman' beings, how would they look? Can we imagine their body? How would it be to be one? What would posthuman experience be like? Can we imagine what they would see? Can we imagine their joy? Can we imagine their *pain*?

My answer is similar to my response to the problem of moral change. First, we could limit our discussion to the near future and use our contemporary values and value feeling to evaluate visions of our future aesthetics and sense. Second, we could try to use and train our aesthetic imagination and taste, if not by imagining future humans and their worlds ourselves (vision again), then by reading or watching good science fiction in order to develop our *taste* for the technological future and the vulnerabilities it might involve. For example, if we think most science fiction is *ugly*, what kind of humans and what kind of world *would* we want then? Aesthetic judgment, like moral judgment, needs to be exercised; we need such exercises for the present as well as for the future. Third, (and this also pertains to ethics), in the design of new humans and a new world, we must try to keep options open for the next generations. In so far as the future can be designed at all, we want to give the next-generation opportunities to co-shape the future world, the future human, the future sense and sensibilities, and the future vulnerabilities and risks, that is, to shape their *world*. This is a matter of *intergenerational* (*ethical-aesthetical*) *justice*.

Fifth, we must further question the relation between normative ethics and normative aesthetics. In particular, we must further question a strict distinction between the two. I already mentioned the *sensus communis* (Kant, Arendt). But there are more ways in which we can show *intrinsic* connections between the ethics and aesthetics of vulnerability. First, normative ethics broadly conceived is not restricted to the moral (questions about the right) but is also about 'the good life' (questions about the good). The issue of the good life has at least an aesthetical aspect. Beauty—in ourselves and in the world—is part of human flourishing, human excellence in the following way. Developing yourself as a human being and having a good life requires developing your senses and your aesthetic judgment. Forming yourself and being formed means (among other things) learning to recognise beauty and becoming more beautiful—becoming a more beautiful person, a more beautiful character—on the way. Ethical training in this broad sense involves aesthetical training.

One could link ethics and aesthetics in an even stronger way and compare the ethical life with the life of the artist. Some think that ethics is a matter of gaining ethical knowledge in the sense of knowing-that. For example, one might think that it is mainly about knowing moral principles. But if we adopt a different, less Platonic and less modern moral epistemology, inspired by Aristotle, Dewey, and Heidegger rather than Plato or Kant, we can think of the ethical life as requiring not knowing-that but knowing-*how*. Then ethics is a kind of *techne*̀, a kind of craft. In order to live the good life, then we need practical knowledge rather than theoretical knowledge. We need to train and practice in order to become better persons.

This view of ethics has consequences for the ethics of vulnerability. If living the good life is an art, then part of what we have to practice to achieve human excellence is to deal with our vulnerability in a beautiful and good way. Coping with your personal and societal vulnerability can be considered an art and a skill. We could try to become 'vulnerability artists', persons who succeed in living a life that is beautiful and good in the way it deals with (reflects on, transforms, etc.) vulnerability—one's own vulnerability and that of others.[5] Then we can raise a new set of ethical-aesthetical questions with regard to vulnerability. Can we cope with our vulnerability in such a way that beauty and good emerge for ourselves and for others?

[5] Note also the connection with post-Aristotelian 'good life' ethics. For example, Epicureans and Stoics had their view on how we should cope with our vulnerability.

Can we let that happen *in spite of* our vulnerability and that of others? Or can we experience it *in* the process of struggling with vulnerability? Can we discern sparkles of beauty and good in the long night of experiencing the possibility of violence and death and in the darkness of existential despair, or more positively, ask from an engaged, active, and practical point of view rather than that of the observer or the bystander: can we be vulnerable in an artful and skilful way?

Finally, with regard to the more concrete question of how the human should look (see above), we could ask if humans should be the measure of all things, as ancient Greeks and Renaissance thinkers claimed about morality (Greeks) or beauty (Renaissance). We can observe, for instance, that today, we tend to build *humanoid* robots. Humans are still the measure. But if we were to become very different kinds of beings, it is likely that the measure will change as well. Perhaps in the long run, it turns out that being human is a fashion, a style? It certainly is that too, among other things. It is not 'just' a style or 'merely' a style. But what is style? In so far as we humans are now able to reshape the human form, style is not 'merely' aesthetical or 'merely' a matter of 'taste'; it is about the kind of form we want to have, about the way we want to say things and do things—the way we want to be-vulnerable. We share with other humans a way of doing, a human style, which is different from the style of other beings (e.g. particular animals or robots). And this style has normative significance: it matters, ethically speaking, *how* we do things as humans. In this sense too, the ethics and aesthetics of vulnerability merge.

Of course, this shared human life form does not exclude differences between cultures. There are also differences *within* one culture and between people. There are various styles of coping with vulnerability available in our culture(s). Some will emphasise their agency and stress that they are in control. Some may fashion themselves as a potential victim or scapegoat. Some will try to follow the stream of the river. And again, some may act as a vulnerability artist or craftsman, who masters the *technè* of vulnerability. Some will accept their vulnerability; others will design various kinds of shields. Some think they are, or could be, invulnerable. Of course, there are other potential attitudes in between these extremes. Vulnerability is one of those places where ethics and imagination meet, and we can try to envision and experiment with alternative ways of coping with vulnerability. Moreover, finding one's style of coping with vulnerability is not done in isolation from others (there are models and anti-models; there is training, disciplining, education), and there is a sense in which communities, societies, and states have to find their style—for example, find their way of coping with the possibility of a (new) nuclear disaster, with (higher) financial risk, or with vulnerabilities generated by new information technologies. Note also, finally, that some people are more influential and perhaps more powerful (the political dimension again) when it comes to projecting a particular 'vulnerability style' as a *better* one than that of others.

In the next subsections, I will explore some answers to the question of *how* to cope with vulnerability and risk, understood as a normative-anthropological project that is as much about ethics and politics as it is about aesthetics and meaning. Which vulnerability transformations and which ways of coping with them are good? Which create beauty? Which do make *sense*?

9.2.1 Fate, Technology, and Modernity: From Greek Tragedy to Cyberpunk

Coping with vulnerability and risk means (among other things) shaping your relation to that which escapes control and agency. In modernity, we emphasise control and try to bring the 'wilderness' of non-control within the domain of our agency. We try to colonise the forests of ignorance and attempt to cross the seas of uncertainty. We transform, manage, and change. We have always done this to some extent—in order to survive, we always had to achieve *some* control over our environment, even if that involved 'low tech'—and perhaps we have always somehow recognised the limits of control and agency. We had to: we run into the wall of reality if we act as if we are invulnerable. Icarus falls when he flies too close to the sun. However, as the last image suggests, cultures deal differently with vulnerability. The ancient Greek attitude towards risk and vulnerability was different from our modern attitude. Is vulnerability a matter of fate, or should we try to overcome it, as transhumanists propose? Is it a form of *hubris* and arrogance to try to overcome it, or is this attempt to overcome it an unavoidable part of human existence?

The ancient Greeks used the term *hubris* in ancient tragedy for actions that challenged the gods and their laws, which resulted in the fall of the main character as the punishment for *hubris*. This idea reflects a particular attitude towards what cannot be controlled by humans. Ancient tragedy shows how human beings are in the hands of fate. They were supposed to respect the laws of the gods and the laws of men, and they had little control over the course of their lives. They felt they had a destiny. This did not mean that humans were entirely passive and had no control whatsoever over their environment. Ancient tragedy was not (entirely) *fatalist*. If we were entirely in the hands of fate, *hubris* would be impossible.

In 'The Ancient Tragical Motif as Reflected in the Modern' (1843) Kierkegaard argued that the 'action in Greek tragedy is intermediate between activity and passivity (action and suffering)' (Kierkegaard 1843, p. 117). Of course, the characters rested 'in the substantial categories, of state, family, and destiny'—this is indeed the 'fatalistic' element in Greek tragedy (Kierkegaard 1843, p. 116)—but the heroes were not entirely passive; they actively contributed to what happened. Thus, the tragic lies 'between' activity and passivity. If we were entirely active and in full control of our lives, there would be no tragedy. But if we were entirely passive, tragic struggle would be also missing. We are neither pure 'agents' nor pure 'patients'. Tragic human action occurs 'in-between' activity and passivity (see also Coeckelbergh 2002).

For coping with vulnerability, this tragic understanding of human existence implies that we recognise the limits of agency when we try to reduce our vulnerability. Often bad things happen to us, and often we cannot do much about them, or we must even recognise that we have *contributed* to it. (It is in his attempt to avoid his fate that Oedipus killed his father and thus did what was predicted.) Tragically, with and in our struggle against suffering, we also contribute to risk and vulnerability. Recognising this does not imply that we should stop struggling or stop doing—this is impossible for us as tragic beings. We are bound to act and live, and in this acting and living, we render ourselves vulnerable. Rather, we must recognise our intrinsic existential vulnerability *in spite of* and partly *because of* our anti-vulnerability actions.

Consider modern technology. At first sight, it seems the summit of un-tragedy since it attempts to gain full control by denying the tragic and vulnerable character of our existence. But in so far as it tries, it never succeeds, and it never *can* succeed. As De Mul has argued (2006), technology is a way of trying to 'tame' fate, but at the same time, it also confronts us with the limits of what we can do and thus paradoxically gives us a sense of the tragic. Similarly, it is not only in our experiences of vulnerability but also in our *anti*-vulnerability struggles *and in the tension between the two* that we meet the tragic. We try to tame technological risk, but risk does not let itself be tamed. It remains wild, and we cannot simply *shoot* it. We have to be patient.

But how patient can we moderns be? In spite of the continuous presence of the tragic in human existence, today we seem to have become less *sensitive* to the tragic. As Kierkegaard also shows, in modernity we put much more emphasis on activity and *individuality*. We can see this in modern fiction. The Hollywood *action* hero is often a strong individual (usually a man) without many ties to others. He is a loner, but not lonely. He does not really need others. Others need *him*: he has to help others. Of course things happen to him, but the message is that he can cope with that—perhaps that we all can cope with it if we really want to. This does not mean that we make the modern hero invulnerable, but he is certainly made less vulnerable than the average human being by his actions. His life is often at risk, but he can deal with that risk. He is in *control*. He controls his vulnerability, and he can take away the risk. There is a threat, but he can save the world. The message seems to be: we can all save the world if we *choose* and *act*. 'Yes, we can' is not only a famous presidential message but also the main message of modern technology and modern culture. We think we can control both nature and technology. In overestimating the power and strength of our wings, we forget that we can fall. We are Icarus. We lack a balance between activity and passivity; activity dominates (Coeckelbergh 2002).

Moreover, the modern hero is not supposed to cope with risk in an emotional way. Like modern politicians, employees, friends, etc., he should not show how vulnerable he (really) is. If he suffers, he should suffer without showing signs of weakness. His strength partly depends on having full control over his emotions. Emotions are a 'no go' for the modern hero: they would show his vulnerability, to others and to himself. Emotions themselves become a source of risk and vulnerability: the hero fears the tears. Like the Stoics, the hero tries to detach himself from anything external, including people. The action hero (usually a man) does not commit to long engagements with either things or people. He does not fear the gods and seeks invulnerability by means of *apatheia*. He has sex but avoids real passion.

When it comes to visions of the technological future, this individualist, Stoic hero, is sometimes placed in a social and material world that is rather hostile and dark. (Here we meet Gnostic dualism again.) Society collapses or is about to collapse. The future is also represented as such, for example, in classic cyberpunk literature. The 'hero' (Is he still a hero?) is a loner who fights for his freedom and his life. The setting is pre-apocalyptic, apocalyptic, or postapocalyptic. The narrative relates to apocalypse since it assumes that we live in an end time and that technology is about to take over. Some transhumanists also hold this view. Kurzweil said: the singularity is near (Kurzweil 2005). In such a vision, there is no place for the (old-style) human. It would mean the end of human life as we know it. It is at the same time the end of humanity and the end of *technology* as we know it.

It seems that influential currents in late-modern fiction—especially those that take seriously the influence of technology on our lives—oscillate between absolute agency (the invulnerable hero) and absolute fate (the singularity or at least the coming of the singularity), thereby denying the existence, significance, and value of a space 'in between', which is the space where all real human life unfolds and where we have to live as beings-at-risk.

9.2.2 Space and Vulnerability: Anti-vulnerability Architectures and Designs

The individualist and dualist world view sketched at the end of the previous section should not surprise us once we consider late-modern anti-vulnerability architectures and anti-vulnerability designs. There has always been such a relation between human-shaped space and vulnerability (e.g. villages can be regarded as anti-vulnerability architectures, and humans try to protect themselves from disasters; see also my discussion of the geography of vulnerability in Chap. 3), but in late-modern times, this relation takes on a more extreme, razor-sharp character since our anti-vulnerability architectures become themselves sources of risk. Let me explain this.

Modernism created (and continues to create) new architectures. This may be regarded as an outgrowth of Enlightenment rationalism. We seek rational control of the world, of the landscape, and of space. But when it comes to our thinking and practices concerning beauty, security, and a lot of other things we value, we remain deeply Romantic. We built modern cities; they are the face of rational modernity. Yet as Romantics, we feel threatened by its modernist social institutions and modernist architectures, which were *meant* to be artificial and make us less dependent on nature by giving us control over nature. They were *meant* to be comfortable and clean palaces that keep out the jungle and secure retreats against the wild—the wilderness of the jungle but also the wilderness in ourselves. They were *meant* to control risks related to people (living together) and threatening environments. But we feel that our individuality has been crushed by these modern structures and we do not feel at home in them.

We still want control and security, of course. But now the modern city is often experienced as an ugly and dangerous place, a place of *risk*. It has become an artificial jungle, an artificial wild. If we go there, we go because we 'have' to work in the city. But we do not really like to *live* there. The environments it creates are workable but not inhabitable and certainly not hospitable. The 'losers' and—worse—the people who *have nothing to lose* live in the city. Others are *commuters*, changers of place. But the place we really like is our homes and (if necessary) our offices—homes. We construct anti-vulnerability cocoons. We create islands of beauty. Offices are controlled spaces. Controlled people live in controlled environments. We work, buy, and consume in places for which Sloterdijk has used the metaphor of the 'crystal palace' (Sloterdijk 2005): comfort zones for the winners of capitalism that excludes others, a sphere of public-private consumption, and living machines for people who seek immunity. A typical example is the shopping centre. But since the city is perceived as a terrible place, we prefer to leave after shopping. We moved out of the city. We build our homes and shopping centres in the suburbs or—if possible—in villages and in the countryside. The home is an island in the traffic sea, a designed oasis in the alien urban and suburban desert. The home is your own fortress. You do not hear the traffic, the thunder of the metro, and the many voices. You do not smell the beggars. You do not smell the risk. You dwell in the light. You feel less vulnerable in there. Moreover, like cloths, the home is not only a protection but is also regarded as an expression of your individuality (very important in Romanticism). There you can be what you really are. Outdoors, in the city, you are part of what Heidegger called the 'they'. You are an employee, a prostitute. You are not yourself. In your home, by contrast, you do things *your* way.

Individuality is not the same as privacy. In late-modern times, we can no longer understand the home as (entirely) 'private'. As I mentioned before, in *The Human Condition* (1958), Arendt still attempted to reinforce the ancient distinction between the private household (the *topos* of dependency and slavery) and the public city (the *topos* of freedom and political action). But today, this distinction is blurred. The city is not very public at all, let alone a place of political freedom; it is a place for economy (household) and business. The office is increasingly a 'private' environment where we meet our friends and where we use online social networks. And as we live our online lives (our 'onlives'), our home lives become increasingly more 'public'. 'Physically' we are at home, but our mind wanders in cyberspace, and when we are online (e.g. in social networks), we are exposed to the cyber public. Moreover, we work at home, and we 'home in' at work. Both spaces—office and home—are 'hybrid' spaces (public and private) and spaces where we try to gain freedom from the alien, modern world.

However, the home and the office are not the only public-private islands where we try to create beauty and where we try to become free. A notorious anti-vulnerability cocoon is the late-modern car, which is designed as a mobile fortress where we can retreat and relax, safe and secure from external threats. It is a shield against the outside world, which is perceived as dangerous and ugly. Cars are beautiful and they protect us. In the car we are in control. We use them as immunity bubbles to move between our palaces (in the city and in the country side). Airplanes and airports are also designed for this purpose. Safety and security are no longer secondary matters; one of the primary functions of air travel is now not only to get from A to B but also to be safe and secure. Airplanes are anti-vulnerability cabins, the ultimate controlled environments where we can feel safe and secure. Everything is done to make sure that comfort and security is maximised—at least for those who can afford it.

Late-modern commuters try to avoid the city centre. If they nevertheless have to go into the (public spaces of the) city, they try to get out as soon as you can. And if they have to be there (for work, for business), then they make sure they are *wired*. Together with other subway cyborgs, they try to make their way through the city. Listening to music allows them to dissociate themselves from the ugly and boring (since it is usually not augmented) environment. The electronic cocoon also allows them to dissociate from others. In order to become less socially vulnerable, they put up a mask, which tells others: 'Do not disturb me, do not talk to me, do not touch me, and do not look at me'. Their invisible immunity shield allows them to glide without much friction through the masses and through the city. Their security bubble is meant to protect them. They try to avoid a *common* world. The prescription is, 'Relate to *your* world. You have your network, your friends, your connections, your experience, and your opinion. In this 'onlife' palace—which is virtual and real at the same time—you can feel secure and at home. Do not let in anything or anyone that may disturb you'.

In spite of all these new anti-vulnerability architectures, however, we remain very vulnerable. Our vulnerability has been transformed, but it is still there. The modern solutions have created their own new vulnerabilities and so have the late-modern solutions. Surely, in late modernity, we no longer feel crushed by 'the system'. We learned to neglect the ugliness of the city. We regained our individuality. We no longer feel that we are a cog in the machine, or rather, we don't mind being a cog in the machine; we love the machine. This change was possible since the nature of the 'machine' changed. They are no longer the large, boring industrial 'machines' we hated. The new machines are exciting *devices*, which are made to be *loved*. They are *beautiful*. We feel that the new 'machines' empower us and liberated us. We gained new 'onlife' lives and worlds. I am not told what to do; I can go where *I like* and see what we *I like*.

But the very measures that were meant to make us less vulnerable to the dangers of 'the system' and 'the masses' create new vulnerabilities. As we move about in our own 'safe room' or 'panic room' and disconnect from our 'nearest' social, material, and natural environment by locking ourselves up in our 'onlife' palaces and comfortable *exclusive* lounges, (1) we also become very *dependent* on our security cocoons and our desire-managing devices (when they fail we are lost: we are violated and we are frustrated), (2) we suffer from new or increased vulnerabilities as a result of the anti-vulnerability technologies we use (e.g. the risks of car travel) and from other forms of ugliness (the home becomes a bunker, the car becomes a tank, I become the ugly extension of my beautiful electronic device), and (3) to the extent that we close ourselves off from our social and natural environment, we risk to miss out on the benefits of having a kind of sociality that is *not* restricted to '*my*world', in which everything turns around me—the centre that stops being a centre when it encompasses everything.

In other words, our 'society' (if that still means anything at all under these conditions) suffers from the absence of a condition of possibility of politics. Politics presupposes a common world. If we have only 'family politics' within our own network, those who are unlucky enough not to be part of many 'families' or networks—those with few 'friends'—cannot make political claims, for example, with regard to freedom or justice. Political values become something I *like* or don't like. And worse, we no longer experience what happens to people outside 'the family'. These people are not part of the face book; in a sense, *they have no face*. Having no face means being out. It also means not appearing vulnerable. And those who do not appear as vulnerable can and will be violated. Those without an account are not taken into account. They are not heard because they do not appear as having a voice, that is, as being vulnerable. They only appear as text or as image-without-face.

In sum, in late modernity, we have intensified our efforts to reach security and to diminish our vulnerability, but this has resulted in new forms of vulnerability, in response to which we have no answer (yet). In the meantime, our shields become unbearably heavy, and the fences we erected become all-inclusive prisons in which we try to expel our fears by means of more distractions. We have become the slaves we never wanted to be. But sadly enough, we are slaves without masters. We cannot blame anyone in particular. We cannot even revolt since we cannot revolt against ourselves. We can only say what we like and what we don't like. We don't want to tell others what to like. At the same time, we become dependent on large corporations, which create architectures for babies. Consumers live in childcare facilities. They are pampered but not educated. Education involves having contact with reality, experiencing resistance to one's desires, becoming aware of risk and vulnerability, and struggling against it or reflecting on it. But we try to avoid the struggle and the reflection—that would spoil the *fun*.

9.2.3 Uncanny Mirrors: Animals, Cyborgs, and Robots

When we reflect on the human form or explore a new, better form of the human, we can take inspiration from non-humans. In fact, we have always done so. To put it stronger, we have always defined the human in relation to non-humans. We have talked about gods, angels, demons, golems, homunculi, machines, monsters, artificial intelligence, zombies, aliens, and so on. We think about the human (what it is and what it should be) by exploring different kind of entities and *forms* of life.

In the West, this descriptive and normative hermeneutical-anthropological exercise has taken the form of what we may call 'negative anthropologies' (similar to negative theology), which define the human in terms of what it is *not* or what it *should not be* (See also Chaps. 4 and 6). For example, we have often referred to 'animals' in order to stress that we are *not* (mere) animals. (See, for instance, Descartes's definition of the human as *not* a beast machine.) We have defined ourselves as political animals, animals with reason, and so on. We use a kind of 'via negativa' to define ourselves. Technologies are also used for this purpose. In particular, artefacts function as hermeneutic devices: they are not only useful but also *meaning*ful; they mediate our perception of the world and of ourselves as humans—in the West and elsewhere—and thus help to define the human.[6]

---

[6] Kaplan has argued that in the West technology seems 'fundamental for defining what humans are' (Kaplan 2004), but we can generalise this to all cultures—which does not mean that there are no cultural differences with regard to this issue. As said, in the West, we seem to have chosen the 'via negativa', but it seems that there are other cultures, for example, in the East, which have different anthropologies.

An example of a technology that is used as a 'hermeneutic' tool within a via negativa towards human self-definition is 'the machine' and, in particular, the robot. In the West, we typically define ourselves as non-machines and (more recently) as non-robots. We do not want to be machines, and we do not want to be robots. Descartes already argued that animals are complex machines but stressed that humans have distinctive features. We may have animal bodies—that is, *robot bodies*—but we also have a soul (Descartes 1637). We are non-machines or more-than-machines. Today we might also make other distinctions and use other terms, but the discourse about the human is still very much influenced by 'the machine question': developments in the field of artificial intelligence and robotics raise questions about machines but also about what we are, about what the human is. Many people define humans in relation to these devices: some compare the brain to a computer; others insist that they do not want to be 'mere' machines or robots and that humans should not be treated as such.

Interestingly, the more artificially intelligent entities resemble us, the better they function as hermeneutic devices. Similarities press us to define our human distinctiveness (or indeed to give up that project). For example, when it comes to human form, it is interesting to observe how robots that *look* like us (that have a similar form) invite strong public responses. The closer they are to us, the more *uncanny* we feel they are. Consider Mori's famous 'uncanny valley' hypothesis: when robots look almost human, they invite an uncanny feeling (Mori 1970). They feel uncanny since they threaten our self-definition. They mirror us, but not entirely. We fear that mirror. It seems that we are anxious to stress that we are different from entities (fictional or not) that resemble the human form, and when this difference is questioned, we feel uneasy.

But if we believe posthuman visions, worse is still to come: what happens if 'the machine' invades me, if I (partly) become a machine? For this particular kind of normative anthropology (Western *negative* anthropology), the ultimate 'monster' is the cyborg: a mixture of human and machine that is monstrous since it crosses the culturally accepted categories 'human' and 'machine' and questions the absolute distinction between the two. It challenges those who wish to keep the human form 'pure'; it offends those who work hard to *purify* the human form from contamination by machine elements.

However, one might also take an alternative view, according to which humans are deeply related to technology—even if their bodies are not 'directly' extended or connected to technological devices. This does not imply that 'anything goes' with regard to human enhancement but rather that the ethical evaluation does not rest on Western negative anthropology and on the related natural-artificial distinction. Then robots and cyborgs can be seen in a less pessimistic and threatening way. Then they can be seen and evaluated for what they *do* rather than for what they 'are'. In order to develop this alternative anthropology, it is interesting to compare Western to Eastern views of technology (e.g. robots) and of the world (e.g. religious views). For example, it seems that in Japan, people are much less worried about ontological distinctions between humans and robots. More generally, it may be instructive to study the more relational and non-dualistic 'ontologies' and anthropologies offered by Eastern thinking (including Chinese philosophies).

If we were to adopt such a non-dualistic view, would it imply that we (have to) give up all (anthropological) distinctions? Not necessarily. Even if we question traditional and modern Western normative anthropologies, which neatly separate humans and machines and actively try to *keep them separated*, we can still make a distinction between different kinds of entities in terms of their vulnerability 'mode' and the way they cope with their vulnerability.

First, even if we question views that make a too strict distinction between humans and non-humans and between biological and artificial, we can and must distinguish between different kinds of vulnerabilities and vulnerability configurations. For example, if we share with (non-human) animals a biological body, then this means we also share a certain kind of vulnerability which is different from the vulnerability of a robot (at least those we know and can foresee today). In this sense, some non-humans are better 'vulnerability mirrors' than others (Coeckelbergh 2010).

Second, like any other entity, humans have a specific way to cope with their vulnerability. For example, they typically develop and use a lot of technology to transform their vulnerability. And last but not least, as outstanding animals (the Heideggerian term is *ek-sistence*), we can *become aware* of our vulnerability and reflect on it. This existential condition creates additional vulnerabilities (see Chap. 3) but also gives us a much larger range of possibilities to relate to our vulnerability than any other entity we know. We can imagine, and experiment with, different ways of coping with our vulnerability, and we can even imagine different vulnerability modes, for example, the kind of vulnerability that might emerge from a 'post-human' existence. The reflections offered in this book presuppose this kind of existential freedom.

Note, however, that even then, these explorations cannot be disconnected from our relations to other, non-human entities. When we try to discern new forms of the human on the horizon, we watch through the glasses of our culture. In the West, this usually implies being afraid of visions that make us (look like) machines. The problem we have with imagery that suggests that 'the machines are coming' means not so much that such robots and (other) artificially intelligent entities may threaten us physically and perhaps enslave us but that they may challenge our idea of the human. We feel that human being is at risk in the sense that we feel that our 'ontological superiority' is at risk.

9.3 Beyond the Human: Environmental and Ecological Vulnerability

If our thinking about the human form is crucially dependent on our thinking about non-humans—that is, if even the Western idea that humans are the measure of all things involves a kind of 'negative' cultural mirroring which implies that *things are also the measure of all humans*—and if what I said about the ecology of vulnerability at the end of the previous chapter makes sense, then asking about the looks, styles, and art of human vulnerability from a more relational and ecological point of view raises questions about the possibility of a *meta-human* aesthetics of vulnerability, an aesthetics of vulnerability that goes beyond the human since it is concerned with both humans and non-humans. Although it seems to me that such an aesthetics will always have to 'depart' from the human subject (we are still the knowing subject) and in this sense remain anthropocentric, we can conceive of several ways in which such an aesthetics could go 'beyond' the human.

First, if the human form depends on the form of those entities we are related to, indeed if the human form depends on the form of the *whole*, then the scope of such a relational and holistic aesthetics of vulnerability goes beyond the human. If we want to think about what kind of form humans should have, we must simultaneously think about the forms of the non-humans and the environments we live in. There is no human form in itself—not epistemologically, since we always interpret the human and have no unmediated access to 'the human in itself' if such a thing existed at all, but also not ontologically: the human form cannot be defined apart from the form of non-humans and the form of the whole.

Second, as a practice, such an aesthetics also goes 'beyond' the human in the sense that human agency is limited with regard to shaping the human form: since it does not control the form of the whole, it cannot (fully) control the human form. The form of the human depends on the form of the whole, on the form of the ecology. The human form is not (only) made; it also grows, and its growth depends on the growth of the whole. Therefore, both human good and human beauty are not entirely in our hands.

Third, as I already suggested, we cannot answer these questions from a detached, 'objective' point of view, and we have to start from the *human* knowing subject, yet this kind of weak *epistemological anthropocentrism* with regard to the aesthetics of vulnerability does not imply that our normative-aesthetical exercise can concern itself with 'humans only'—if this makes sense at all. From a relational-existential point of view, in the end, the question concerning a normative aesthetics of vulnerability cannot be answered outside lived experience and action where humans and their environment meet, where mind and body, spirit and matter, and human and world dance. The form of human being and the form of existential vulnerability emerge *in* that dance. In this sense, we must use a non-anthropocentric epistemology or rather, an epistemology that goes beyond the anthropocentric/non-anthropocentric distinction and defines knowledge as emerging from engagement and *worlding*. We are already related with non-humans, and these relations shape and constrain our being-vulnerable and our thinking about our own vulnerability. A relational-ecological conception of existential vulnerability understands vulnerability experience and vulnerability struggle in a way that firmly links us with non-humans, with the world. Both vulnerability knowledge and vulnerability action go beyond the human. The *forms* of our vulnerability struggles are deeply related to the forms of other, non-human struggles. Humans are only co-authors with regard to how they cope with vulnerability and how they shape themselves; these stories are written by many 'authors'.

To conclude, if existential vulnerability and human being are intrinsically ecological, then their beauty (and their good) refers us beyond the human, and the 'human' song depends on many non-human voices. How long this song continues, how it will continue, and whether or not a new kind of music emerges are only partly up to us. Our practices of normative anthropology are tragic in this sense. We might try to enhance ourselves; we might even attempt to achieve a posthuman mode of existence. But we do not know the future of the human and the future of the human form—if there will still be humans at all. We can try hard to shape that future. Perhaps we want something new. But with some Nietzschean courage, we may as well say da capo!

References

Arendt, Hannah. 1958. *The human condition*. Chicago: The University of Chicago Press.

Coeckelbergh, Mark. 2002. *Liberation and passion*. Alfter/Bonn: Denkmal Verlag.

Coeckelbergh, Mark. 2010. Artificial companions: Empathy and vulnerability mirroring in Human-Robot relations. *Studies in Ethics, Law, and Technology* 4(3). Available at http://www.bepress.com/selt/vol4/iss3/art2/

De Mul, Jos. 2006. *De domesticatie van het noodlot: De wedergeboorte van de tragedie uit de geest van de technologie*. Kampen: Klement.

Descartes, René. 1637. *Discours de la méthode*. Trans. L.J. Lafleur. *Discourse on method*. In *Discourse on method and meditations*. Indianapolis: Bobbs-Merrill, 1960.

Kaplan, Frédéric. 2004. Who is afraid of the humanoid? Investigating cultural differences in the acceptance of Robots. *International Journal of Humanoid Robotics* 1(3): 465–480.

Kierkegaard, Søren. 1843. *Either/or: A fragment of life*, vol. 1. Trans. D.F. Swenson and L.M. Swenson. Princeton: Princeton University Press, 1944.

Kurzweil, Ray. 2005. *The singularity is near: When humans transcend biology*. New York: Viking/Penguin.

Mori, Masahiro. 1970. The uncanny valley. Trans. K.F. MacDorman and T. Minato. *Energy* 7(4):33–35.

Sloterdijk, Peter. 2005. *Im Weltinnenraum des Kapitals*. Frankfurt am Main: Suhrkamp Verlag.

Part III
Conclusion

Chapter 10
Conclusion: The Heel and the Arrow

Let me briefly summarise how these reflections have contributed to discussions about risk and vulnerability, to the ethics and politics of human enhancement and information technology, and to contemporary philosophical anthropology.

First, by understanding risk and vulnerability as terms that tell us something about the existential relation between subject and object, I have shown how we can avoid both objectivist and subjectivist approaches to risk and vulnerability—without giving up some valuable insights about the human, experiential, contextual, and environmental dimensions of the concepts of risk and vulnerability. Risk and vulnerability are about the kind of *relation* we have to the world as human beings, and objectifying or psychologising them denies their relational and existential character.

Second, by shifting the focus from 'human nature' to 'human being' and human existence and therefore from a static to a variable but non-naturalistic (pre)conception of what we are, I have helped to open up a third route between scientistic-materialist, naturalist, and transhumanist progressivism, on the one hand, and metaphysical and anthropological conservatism, on the other hand. Importantly, this shift has been managed without giving up a normative discussion. I have argued that we can and must accept anthropological change at the descriptive level (we must accept *that* human being changes) but that at the same time, we can and should discuss *which* changes we want—at least to the extent that we have the power to make these changes and to the extent that we can imagine and sense future possibilities.

Third, in this book, I have restricted this descriptive and normative discussion to the theme of vulnerability and risk. However, to discuss ethics of human enhancement in these terms could easily lead to a new polarised discussion: one between transhumanists who want to dispose of human vulnerability by means of human enhancement and opponents who assert and value the intrinsic vulnerable nature of humans and use that as an argument against human enhancement (e.g. neo-Aristotelian, ethics of care, or feminist opponents). Similarly, this could generate a discussion between those who wish to transcend human vulnerability by means of information technology and those who defend the value of vulnerability and embodiment and see Internet and other technologies as threatening to vulnerability and to human values. Instead,

M. Coeckelbergh, *Human Being @ Risk: Enhancement, Technology, and the Evaluation of Vulnerability Transformations*, Philosophy of Engineering and Technology 12,
DOI 10.1007/978-94-007-6025-7_10, © Springer Science+Business Media Dordrecht 2013

I shifted my discussion to new and better questions: a descriptive question about the history and future of vulnerability transformations and a normative question about which of such ethical-anthropological, political-anthropological, and aesthetic-anthropological changes we want. Neither invulnerability nor social-technological paralysis is an option for us as always-already-vulnerable beings, whose vulnerability is always transformed by technology, which changes together with the social. The question 'In so far as we have the power to influence the form of the human, what do we want to become?' always implies the question, 'In so far as we can know and influence the vulnerability transformations that will happen in the future, what (kind of) risks and vulnerabilities do we want to experience, create and foster, and how do we want to cope with them?' and therefore 'What kind of technologies do we want?' and 'What kind of society do we want?'.

Here is another, more metaphorical way of formulating that normative challenge. An ancient Greek myth tells us the story of Achilles, who was made invulnerable in his youth except his heel. He is said to have died from a heel wound caused by an arrow shot at him. In this book, I have argued that posthumans would not be unlike Achilles in this respect. If and in so far as posthuman heroes and their creators try to transcend vulnerable existence, they are bound to fail because they have—*and always will have*—heels. Of course, some kinds of vulnerability may be diminished in the future. But as we create and use new technologies, new heels are created. This book suggests that if we can and must make an ethical choice at all, then that is not a choice between vulnerable humans and invulnerable posthumans, or between vulnerability and invulnerability, but a choice between different *forms* of the human and different *forms* of vulnerability. Moreover, if risk is not external, then we must dispose of the dragon metaphor in this context: vulnerability is not a matter of 'external' dangers that threaten or tyrannise us and that have nothing to do with what we are; instead, vulnerability is bound up with our relational, technological, and temporal kind of being—human or posthuman. If there are 'dragons' at all, they are part of us, and we are part of them. As I put it in a recent article (Coeckelbergh 2011), we are—and we will remain—at once the heel and the arrow, the hero and the dragon.

I conclude that we can and must criticise transhumanist proposals for human enhancement and evaluate new information and communication technologies, but that this should not be done by relying on the notion of a static human nature, on a dualist view of humans and their world, or on objectivist or subjectivist conceptions of risk and vulnerability, but by asking *which vulnerability transformations* we really want. Although we cannot fully control future vulnerability transformations, it is important that we evaluate what we are doing in the light of current and expected future vulnerability-technology configurations.

This exercise should also involve further reflection on the current predominant way of conceptualising and managing risk and vulnerability; on the stretchability of our imaginative and emotional capacities with regard to predicting future value changes and future vulnerability transformations; on the relations between power, society, and vulnerability; on the fairness of present and future distributions of vulnerability; on the art of coping with vulnerability; and on the ecological nature of vulnerability.

Admittedly, this is a lot to ask. This essay offers only preliminary reflections on these issues and is at most a 'prolegomenon' to such a project: its preliminary and provisional nature becomes clear once we realise how little we can say about the future of the human and how far we are still removed from a comprehensive, systematic, and compelling conceptual framework for thinking about human vulnerability that could guide and inspire our experimental, often rather clumsy attempts to find better human-technological possibilities. But a bold and imaginative normative anthropology that prepares us for the future should not be content with less.

Reference

Coeckelbergh, Mark. 2011. Vulnerable cyborgs: Learning to live with our dragons. *Journal of Evolution and Technology* 22(1): 1–9.

Author Index

M. Coeckelbergh, *Human Being @ Risk: Enhancement, Technology, and the Evaluation of Vulnerability Transformations*, Philosophy of Engineering and Technology 12, DOI 10.1007/978-94-007-6025-7, © Springer Science+Business Media Dordrecht 2013

Subject Index

M. Coeckelbergh, *Human Being @ Risk: Enhancement, Technology, and the Evaluation of Vulnerability Transformations*, Philosophy of Engineering and Technology 12, DOI 10.1007/978-94-007-6025-7, © Springer Science+Business Media Dordrecht 2013

Printed by Printforce, the Netherlands